HOT HOUSE

HOT HOUSE

Global Climate Change
and the Human Condition

ROBERT STROM

Copernicus Books
An Imprint of Springer Science+Business Media

In Association with
Praxis Publishing Ltd

PRAXIS

Published in the United States by Copernicus Books,
an imprint of Springer Science+Business Media.

Copernicus Books
Springer Science+Business Media
233 Spring Street
New York, NY 10013
www.springer.com

Library of Congress Control Number:
2007929781

Manufactured in the United States of America.
Printed on acid-free paper.

9 8 7 6 5 4 3 2 1

ISBN 978-0-387-34179-8   e-ISBN 978-0-387-68611-0

*To the world's grandchildren.
May your parents, grandparents and their
policy makers be wise enough to preserve
your future.*

Part of the generation that is depending on us to secure their future. (School children at Takumabaru Elementary School, Kumamoto City, Japan).

CONTENTS

Contents

Contents

Contents

Contents

FOREWORD

Climate change: how bad is it, what's causing it, how bad might it get, and how can we avert the worst and even prosper into the future? Strom summarizes a vast body of scientific data detailing answers to these questions. The bottom line is "serious but survivable" for most of Earth's inhabitants in the best possible outcome, which may be achievable if we have lower than expected climate sensitivity to greenhouse gases and if we restructure our economy to emphasize efficiency and alternatives to fossil fuels. Civilization might thrive under the best scenarios. So there is good news. The worst possible outcome may entail, just as Strom describes it, the collapse of civilization and the demise of many life forms on Earth. That demise could include extinction of humans if anarchy and loss of technology ensue as a consequence of simultaneous severe climatic disruptions and severe resource depletion.

There are gradations between these optimum and worst-case hypothetical outcomes. We don't know which fate Earth will have, or what measures would narrowly avert civilization's collapse versus those that would enable civilization to move smartly into the future. We do know that business as usual is untenable if our grandchildren are to have stable lives as prosperous as human ingenuity and nature can make them.

The trends of accelerating climate change and diminishing resources pose dual dilemmas. The power of the mathematical exponent cannot be overcome unless the mathematical functions change. It is not a matter of whether the two trends, or either of them alone, will cause the implosion of civilization, but a matter of when, unless the premises of our current economic system are broken. A problem is that individuals and nations all expect to consume more stuff next year and next decade than this year, and so on into the indefinite future; most people relate their well-being to an exponential growth in consumption of stuff. In fact, an Efficiency

Revolution coupled with population stabilization could enable improved, expanded human well-being indefinitely into the future. Current economic premises do not allow this.

Scientists often criticize the political leadership for a failure to understand the climate situation and its links to combustion; we point out that to wait another generation for corrective action could be too late. In fact, a generation has already been consumed as scientists debated the evidence. There was no other way, as the warming trend had to emerge from the envelope of natural fluctuation, and responsible science required careful evaluation of evidence. Politicians understandably will require more time to accept that climate scientists had their debate and now are virtually unanimous in recognizing an acceleration of climate change related to combustion. But time is not on our side.

Critics of scientific research into climate change appropriately point out the large uncertainties in climate models. Part of the uncertainty is not inherent in the data or models, but it is embodied in political indecision caused by current obfuscation about the actual climate record and the impacts of climate change. Little has been done so far to change things, at least in the USA and some of the fastest growing nations; not enough is being done, even in the most progressive countries. Even the Kyoto Agreements lack the global comprehensiveness needed to attack the problem and achieve the best for humanity. Thus, the climate models must incorporate a large range of possible greenhouse gas emission scenarios, which compound real uncertainties inherent in state-of-the-art climate modeling. The uncertainties, in fact, skew possibilities toward worst-case scenarios because of the indecision about actions to improve our future. There is nothing else that can be done responsibly but to consider the warnings offered by those worst-case scenarios (along with modest ones).

I personally cannot take the most severe models too seriously; the reason is simply that I cannot imagine intelligent leaders being so blind to real threats that they would remain indefinitely antagonistic to science and resistant to solutions to the overwhelming climate-change problem staring all of us in the face. I am optimistic that the worst scenarios outlined by Strom will not elapse. It is almost not even worth my consideration of them, except for one nagging thought: what if I also err in my optimism that America's and the world's political leadership will get a grip on reality and respond in such a way that they shape the future favorably for the American and global common good? What if logic is not the leadership's guide? What if living for the moment of selfish indulgence is the dominant control on the human course? History would suggest that logic is important in leadership, but self-indulgence has been just as important in the long trendline of history. Which will dominate this generation's thinking and actions?

The uncertainties remaining in climate models are important for scientists and policy makers to acknowledge, understand, and reduce. Equally important is that the public and political leadership recognize the unacceptable consequences of unimpeded climate change. We can debate and model whether civilization could survive at all, struggle through it, or adapt and prosper. The fact is, we don't know. It is so difficult to project the positive impacts of human ingenuity in dealing with adversity, and of sheer luck in staving off the end of humanity as we know it. It makes sense to delve into our condition and our challenges; it makes no sense to argue while the house is burning.

Given that there is an overwhelming resistance to make the changes that are so urgently needed, incredulous questions arise. Why would humanity utilize nearly all the ingenious devices and manufacturing capability brought about by the Scientific Industrial Revolution to increase adversity to civilization? Why would the political establishment view the contributions of modern science so admiringly when it comes to creating the means to improve life and also to destroy the Earth, and then view modern science with such suspicion and antagonism when it comes to studies that point to severe problems and solutions to the problems? My questions may seem defensive. But it is a sense of defending all of us from unmitigated catastrophe that prompts these questions.

This book delves mainly into climate change issues and the need to conserve fossil fuels and to develop alternatives to them; but resource depletion is another problem of almost comparable magnitude. Urgent conservation can substantially solve both problems. A problematic aspect is that any individual person and any individual nation is only fractionally responsible for the problems. Any four consecutive years' of resource waste and carbon emissions only fractionally change the picture. It thus makes sense from a pure matter of electability that most of our politicians advocate throwing energy at economic productivity to give a sense of prosperity. People seem to have an incipient recognition of this problem and of the need to stabilize and reduce consumption. Our politicians should listen to the people, and then start to lead them.

A nexus of severe problems all push for serious and urgent conservation and development of alternative energy sources in the United States and abroad: wars and rumors of wars; defense expenditures to defend petroleum and gas fields and hydrocarbon shipping lanes; the American current account trade deficit and federal budget deficit; American prestige, honor, and influence abroad; climate change impacts on agriculture and coastal communities; air quality; and human health and health sector costs are just a few. If America could start to define and implement solutions, the world would follow. A more comprehensive "Kyoto Plus" accord would be a start.

The nexus of problems would be substantially reduced if oil and fossil fuel consumption was reduced sharply. Had the three-year costs of the Iraq war instead been spread over 12 years of tax incentives and industry incentives to produce and market hybrid vehicles, we could hybridize the entire American fleet of new internal combustion engine-powered vehicles—every new car, bus, bulldozer, and tank. The oil savings would exceed Iraqi oil production, and the cost savings in every sector would propagate. That would be just for starters. Efficiency is a powerful engine of genuine economic growth and prosperity. Waste serves nobody but the sellers of wasted products.

As Strom points out, there may exist a particular threshold of greenhouse gas emissions that may trigger profound, sudden changes to the Earth system. Perhaps more likely, Earth is apt to undergo cascading series of changes triggered by a succession of thresholds, rather than a single doomsday threshold. In other words, we may have multiple opportunities ahead to trigger increasingly more dismal fates, and also multiple opportunities to avert the very worst.

Like Scrooge in Charles Dickens' *A Christmas Carol*, it is likely that people will see their alternative futures, then change their collective behavior in order to choose the best outcome. Formulation of a plan for success will require open discussion—not fear mongering, not sensationalism, and, equally so, not obfuscation and blind refusal to look at cause and effect. It will require discourse among scientists, economists, industrialists, and policy makers to consider novel ways to achieve an efficient, optimum future society.

Dr Jeffrey S. Kargel

Dr Kargel is a research scientist at the University of Arizona, and he is the director of the 28-nation consortium Global Land Ice Measurements from Space. The views presented here are his alone.

PREFACE

I am not a climate scientist or an environmentalist. I do not belong to any environmental organization such as the Sierra Club, the World Wild Life Fund, or Greenpeace, although I am sympathetic to many of their causes.

I am a retired planetary scientist specializing in impact cratering and planetary geology, including catastrophic flooding and implications for short-term climate change on Mars. I have spent my entire career studying global planetary systems, and I have a general understanding of how they work. Although I am retired, I am still active in planetary research and the MESSENGER mission to Mercury. The science stated in the text has mainly been taken from peer-reviewed science journals. It is listed in the References section, which is followed by categorized references under Further Reading.

I became interested in global warming almost 15 years ago when I started discussing it in a planetary geology course and later in an undergraduate seminar called "Planetary Catastrophes." This seminar included discussions of catastrophic events such as impact cratering, catastrophic volcanic resurfacing on Venus and Jupiter's satellite Io, and catastrophic flooding on Mars. I decided to include a discussion of global warming because it has the potential of becoming a catastrophic event for humans and other species if it is not brought under control. It is happening very rapidly even on a human time scale, so certainly on a geologic time scale it is a potentially catastrophic event.

The more I studied the subject, the more fascinated I became with climate change and its effects on the Earth and its biosphere. As a planetary scientist it is particularly interesting to me because global warming affects the entire planet and I relish dealing with the whole planet and even the entire Solar System. When I started discussing this topic 15 years ago, many aspects of the science were still uncertain. At that time it was fairly certain that the Earth was warming, and very certain that the atmospheric carbon dioxide level was increasing rapidly. The human role in global warming, and whether or not global warming was an event primarily due to natural causes, was still being debated. Those issues are no longer debated among scientists, although there are a few who will just not face reality.

Climate change is a very complex, multidisciplinary subject. In this book I have attempted to cover all aspects of the problem in a manner that is understandable to the educated non-scientist. The latter parts of the text deal with some of the political issues and possible actions to mitigate the problem.

Global warming is a rapidly evolving science where new studies are published almost daily. Undoubtedly by the time you read this book important new data will have been published in the science literature. However, the main concepts in this book are unlikely to change significantly and, hopefully, I will publish a revised version in several years. In the meantime, you can keep up-to-date by viewing the websites listed in Appendix D. I hope you find the book interesting and will do all in your power to help solve our problem.

ACKNOWLEDGMENTS

Without the help of Lisa Martin this book would not have been possible. Lisa is the science librarian for the Lunar and Planetary Laboratory at the University of Arizona. She did reference searches, kept her eyes open for articles on global warming and related issues in the science literature, and read a draft of the book to see if I was getting the material across to the non-scientist. I thank her from the bottom of my heart for all the work she did. I especially want to thank Jeff Kargel at the University of Arizona for reading the book and making very helpful additions and corrections. I also greatly thank Dr Philippe Blondel of the Department of Physics, University of Bath, and Dr John Mason, Chief Subject Advisor Editor, Praxis Publishing, United Kingdom, for critically reading the book and making very useful suggestions for its improvement. Dr Harry (J.J.) Blom of Springer Publishing read portions of the manuscript and made very helpful comments to improve the book. My brother Jack was very helpful in suggesting improvements to enable the non-scientist to better understand the material. I also want to thank Jim Torson of the US Geological Survey (retired) who, by almost daily e-mail, provided a wealth of valuable information about climate change. I would also like to thank Clive Horwood of Praxis Publishing for his support throughout and Alex Whyte for editing the manuscript and catching many reference inconsistencies. Finally, I want to thank my wonderful wife Ayako for her encouragement, support, and patience as the book was being written. Without her continual prodding the book might never have been completed. Thanks, darling.

Temperature Scales*

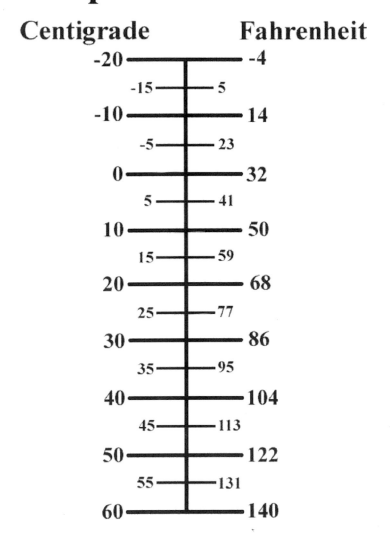

Centigrade   Fahrenheit

Centigrade	Fahrenheit
-20	-4
-15	5
-10	14
-5	23
0	32
5	41
10	50
15	59
20	68
25	77
30	86
35	95
40	104
45	113
50	122
55	131
60	140

*A degree Centigrade spans 1.8 degrees Fahrenheit.

1

A GIGANTIC EXPERIMENT

U nwittingly in the beginning and knowingly at present, we are conducting a gigantic, uncontrolled experiment with the Earth. Although many outcomes of this experiment are uncertain, they will surely be very harmful, if not catastrophic, for humans and many other species. Even if we could stop the experiment now, the forces we have already unleashed will haunt us for centuries. The experiment is global warming. It is due to the human abuse of the environment as a consequence of the Industrial Revolution and the unrestrained population growth during the last 250 years. Although we did not know it then, the experiment began to affect the composition of the atmosphere with forest clearing and coal burning in the Northern Hemisphere during the 17th, 18th, and 19th centuries. With the coming of the Industrial Revolution in 1751, and the advance of technology and science in the 20th century, changes in the climate intensified to the point where they are readily detectable today.

The problem might be described in medical terms. We have been "smoking" (emitting greenhouse gases) for the past 250 years, and now we have "lung cancer" (global warming). Like cancer, global warming can be just as fatal. Fortunately, this "cancer" has been detected in its relatively early stages by the tools of modern science. It is, therefore, "operable" with hopefully only minor side effects, but only if we act now to curb greenhouse gas emissions (stop smoking). If we do not begin the "operation" soon, or if we trigger a reaction that spreads the "disease" very rapidly (abrupt climate change), then it will be too late to head off the severe, possibly fatal, consequences.

Like cancer, global warming is an insidious "disease." When a person first develops cancer there are no symptoms. In the earliest stages of global warming there were also no symptoms. As cancer progresses, the afflicted person may have minor aches and pains, but these could be mistaken for aging or some other benign physical problem. That is roughly where we now stand in the case of global warming. We are just beginning to feel its effects, but so far they are fairly benign and have only recently been recognized as due to global warming. The person may go to a doctor to check on the minor aches and find that he or she has cancer. Of course, this is very difficult to accept, so the person may go to nine other doctors for their opinions. Let us suppose that nine of the 10 doctors diagnose that the person has cancer, and that he or she should have an operation to remove it before it spreads and becomes inoperable. However, one doctor thinks it could be something else and advises the patient to wait and see what develops. What would you do? Most rational people would have the operation, but some would take the lone doctor's advice because it is more reassuring; they are essentially in denial. In denial today are a few scientists and numerous politicians and media people who refuse to look at the evidence for global warming and its causes.

The scientific consensus is that global warming is rapidly occurring and that human activity is primarily its cause by emitting greenhouse gases. They also agree that we can curb the worst effects by significantly reducing the emission of these gases if we act soon. We have been warned, but currently we are not responding to these warnings in any significant way.

A COMPLEX SUBJECT

Global warming is extremely complex because it deals with so many different characteristics of the Earth and has complex interactions. It is addressed by almost all sciences, including many aspects of geosciences (particularly glaciology, hydrology and oceanography), atmospheric sciences (especially atmospheric physics and climatology), the biological sciences (particularly ecology and biochemistry), and even astronomy (solar physics and celestial mechanics). It has recently become the concern of other diverse disciplines such as economics, agriculture, demographics and population statistics, medicine, engineering, and political science. If a person looks at only one or two aspects of the climate change problem, then that person is looking at it with blinders. It must be viewed from a planetary perspective encompassing all the above science disciplines. This book attempts to address these complex interactions, integrate them, and derive meaningful conclusions requiring certain actions by the world community.

MISUNDERSTANDING OF GLOBAL WARMING

There is an enormous misunderstanding of global warming among people not familiar with it, and, unfortunately, that is the vast majority. Although most people have heard of global warming and even accept it, they do not know exactly what it is, how it works, or how it could affect their lives. They are often confused over conflicting reports in the media and statements by politicians. This often leads them to question whether or not global warming is happening. Even a few scientists, most who have not studied the problem or may be influenced by their funding sources, are skeptical of global warming or even believe it is beneficial. They are having less and less influence on the media and policy makers. Most of the public do not realize that there is a very strong scientific consensus among climate scientists that global warming is not only occurring but that humans are largely the cause, and that it poses a grave human threat.

Policy makers and the public are often confused when scientists debate some aspect of science. They hear different interpretations and think that as the science community cannot agree they really cannot know very much about the subject. Furthermore, scientific conclusions are often given statistical probabilities. Even if the probabilities are statistically very high, non-scientists think there is much uncertainty and, therefore, that the science is unreliable. They also fail to realize that most often scientists are debating rather small parts of a much larger principle or concept that is generally accepted by all. As a figurative example, consider a group of scientists that was asked, "Is the Earth round?" Maybe 30% would say

"Yes," another 30% may say "No, it's spherical," still another 30% may say "No, it is an oblate spheroid," and 10% might say "We don't have enough data to determine the exact shape of the Earth." Although they all agree with the general shape, round instead of flat, they may differ on the details. In effect all their answers are correct. That is similar to the situation with global warming; almost all scientists agree that it is happening and that humans are the cause, but some differ on details, as will always be the case with science.

CURRENT UNDERSTANDING OF THE PROBLEM

Most scientists who have delved into the subject agree that global warming is a reality caused primarily by human activities, and that it is a serious problem that must be dealt with now. The vast majority of science organizations around the world endorse the findings of the scientific community that studies climate change and global warming. They have advocated that governments take an active role in curbing human-produced greenhouse gas emissions that are causing the problem. These organizations include the following groups:

- Intergovernmental Panel on Climate Change (IPCC), a United Nations agency made up of over 2,000 scientists from a variety of countries and scientific disciplines.
- US National Academy of Sciences, the most prestigious scientific body in the United States.
- American Geophysical Union, the largest geoscience society in the world.
- Britain's Royal Society
- American Association for the Advancement of Science (See Appendix C)
- American Meteorological Society
- Over 20 other scientific societies scattered around the globe.

The vast majority of climate scientists and scientific organizations agree to the validity, or uncertainty, of various aspects of global warming. This consensus has evolved over the last 25 years. Every 6 years the IPCC issues a comprehensive report evaluating climate change. The IPCC is made up of over 450 lead authors and over 800 contributing authors from over 130 countries. The report is reviewed by over 2500 expert scientists. The latest report is the fourth assessment issued in 2007. It consists of four main parts: Physical Science Basis, released February 2, 2007; Impacts, Adaptation and Vulnerability, released April 6, 2007; Mitigation of Climate Change, released May 4, 2007; Synthesis report, due November 16, 2007. The report was finished at the end 2005 and then circulated to countries for their review and comments during 2006. This book was written before the report was released, but it essentially agrees with almost everything discussed in the report. This is not surprising because the information in the report came from the same sources as this book: the peer reviewed science literature. However, there have been some new developments since the IPCC report was finished at the end of 2005. These are contained in this book. Furthermore, there is no discussion in the IPCC report about the role of population growth on current climate change. That is discussed in Chapter 9. The primary conclusions of the Fourth Assessment are that global warming is "unequivocal", and that the primary cause is the human-caused release of greenhouse gases at a certainty >90%. The conclusions stated in this report are similar to those in the Third Assessment Report issued in March 2001, but the numbers have firmed up and the uncertainties have decreased appreciably. Naturally it takes a fairly long time to build a base of observations to test whether global warming is happening, its causes, and its consequences. At present (2007), the vast majority of scientists agree to the first four general states of our knowledge about various aspects of climate change. In the list below, the percentage probability is given in parentheses. If the percentage is over 90%, you can be assured that it is happening. The last three aspects of global warming are less well known, but the best estimates of the percentage probabilities are given in parentheses. The uncertainties will surely decrease with time.

1. Increase of greenhouse gases—firmly established (100%)
2. Increase in global average temperature—well established (95–100%)
3. Weather indicators of global warming—likely to very likely (66–99%)
4. Cause of present global warming—likely understood (>90%)
5. Consequences of global warming—some well known, but many are uncertain (30–100%)
6. Solution to the problem—well understood (90–99%)
7. Prospects of initiating a solution by industrialized nations—extremely uncertain (0–10%)

All of these aspects of global warming will be discussed in detail in subsequent chapters. The last chapter discusses the pessimistic outlook for item 7.

To test the unanimity of the climate change community on the issue of the reality and human cause of global warming, Naomi Oreskes (2004) analyzed 928 abstracts on climate change listed in the ISI scientific literature database. All the papers were published in peer-reviewed scientific journals between 1993 and 2003. They were divided into categories that represented explicit or implicit endorsement of the consensus position, methods in climate change research, paleoclimate analysis, and rejection of the consensus position. Of the 928 papers, 75% either explicitly or implicitly supported the consensus position and 25% dealt with methods or paleoclimate that took no position on current human climate change. Incredibly, there were no papers in the peer-reviewed scientific journals listed in the ISI database that disagreed with the consensus position.

There are reports from individuals in non-peer-reviewed publications that dispute global warming and its causes. There are also several scientists who maintain that global warming is not happening or that it is just a natural fluctuation in the climate that will eventually return to "normal" conditions. While writing this book no articles were encountered in the science literature written by these scientists refuting the consensus on global warming. However, at least one (Richard S. Lindzen of the Massachusetts Institute of Technology, USA) has been writing his views in such revered publications as the *Wall Street Journal*. If there really is solid scientific evidence that we have nothing to worry about, then such scientists are morally obligated to submit it to the science literature for peer review. If their ideas have any merit, why are these scientists being ignored by their colleagues and why are they not submitting their evidence to science journals?

The public should realize that in every science there have always been deniers of established principles in their fields. One of the most famous is Sir Harold Jeffreys, the very eminent geophysicist and astronomer, who opposed plate tectonics until the day he died in 1989. He was not alone. There were other Earth scientists who also denied plate tectonics after it had been firmly established. The same thing is happening today with regard to global warming. The big differences are that the present deniers do not have anything close to the stature of Sir Harold, and some politicians are using these people to advance their own agendas.

In searching the science literature for evidence, I found one paper by two authors (McIntyre and McKitrick, 2005) that challenged, on statistical grounds, the data from Mann et al. (1998, 1999) that the present warming is unprecedented in the last 1,000 years. This paper maintained that past climate changes during that time were similar to today, and, therefore, today's warming is a natural fluctuation not resulting from human-caused emission of greenhouse gases. However, this paper was strongly criticized by several scientists (Huybers, 2005; Moberg et al., 2005; Osborn and Briffa, 2006), and refuted by at least 13 other papers representing 45 scientists (see Chapters 6 and 7). It was also refuted by studies mentioned in the UN IPCC 2001 and 2007 reports and more recently by the National Academy of Science report by 12 scientists (National Research Council, Committee on Surface Temperature Reconstructions, 2006). Should I have reported on the McIntyre and McKitrick paper and ignored the others? I think not. However, their paper is listed in the references and discussed in a footnote in Chapter 6 (see page 83). If there had been a substantial number of scientists who published in the science literature opposing views to the main conclusions on global warming, their views would have been discussed in this book. In fact, if that had been the case, this book would never have been written.

On June 8, 2005, prior to the July meeting of the G8 nations in Gleneagles, Scotland, the science academies of those and other nations urged their governments to take prompt action to combat global warming. The academies included the US National Academy of Science and the British Royal Society. It was signed by leading scientists from 11 countries, including two of the biggest greenhouse gas emitters, China and India. Their statement read in part: "It is likely that most of the warming in recent decades can be attributed to human activities. The scientific understanding of climate change is now sufficiently clear to justify nations taking prompt action. Action taken now to reduce significantly the build-up of greenhouse gases in the atmosphere will lessen the magnitude and rate of climate change." Science academies of leading nations do not make joint statements about issues that are controversial or in doubt. The fact that they made such a statement about global warming is unprecedented and strongly indicates that global warming is real and that human activities are primarily the cause by releasing greenhouse gases.

Despite this strong endorsement by scientists, their societies and academies, many policy makers and the media continue to assert that climate change science is highly uncertain and that we should not take strong measures to curb greenhouse gas emissions until the subject is better understood. For example, Senator Ted Stevens, Republican from Alaska, who was (2006) the chairman of the Senate Committee on Commerce, Science and Transportation, stated that although he thinks global warming is occurring, he does not accept the conclusion that the driving force is the release of greenhouse gases by human burning of coal, oil, and natural gas. According to his spokeswoman the action that Stevens is taking is to fund more research about climate change. This is an incredible statement from a Senator whose state is falling apart from global warming (see Chapter 10). Another statement by a politician, Senator James Inhofe (Republican from Oklahoma), is absolutely appalling, even bizarre. He has stated, "With all the hysteria, all of the fear, all of the phony science, could it be that man–made global warming is the greatest hoax ever perpetrated on the American people? It sure sounds like it to me." Apparently these Senators reject, or are ignorant of, the approximately 700 peer-reviewed science papers that find global warming to be real and that the cause is primarily the human burning of fossil fuels. This attitude among American policy makers is one of the reasons for assigning only a 0–10% probability of initiating a solution to the problem.

The following chapters attempt to present an unbiased analysis of climate change gleaned from the science literature, particularly during the past six years. Uncertainties in various aspects of the science are discussed and firm conclusions are emphasized. By the end of the book, it should be obvious that we have a serious problem requiring immediate action.

2

THE ABCs OF CLIMATE CHANGE

IT is important to understand certain basic concepts about climate change before we delve into its causes and its consequences. These processes will be incorporated in the topics discussed throughout this book. They are fundamental to understanding the broader aspects of climate change and global warming issues.

THE GREENHOUSE EFFECT

The greenhouse effect is named after greenhouses that let in sunlight to heat the interior and then trap it. The heat is trapped by the glass and warms the interior. Although this analogy is very far from perfect, it does convey the general principle behind the process of greenhouse warming in planetary atmospheres.

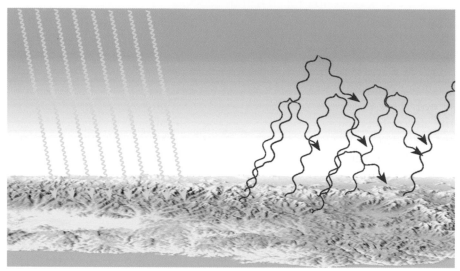

FIGURE 2.1 *This graphic illustrates the greenhouse effect. Sunlight (yellow lines) falls on the ground and warms the surface. The warm surface radiates the heat at infrared wavelengths (red wavy lines). This heat is absorbed by greenhouse gases and reradiated back to the surface. (Courtesy Tasa Graphic Arts, Inc.)*

In the greenhouse effect, solar radiation is absorbed by the Earth's surface resulting in surface warming. Much of this absorbed energy is eventually reradiated at longer infrared wavelengths as heat. A portion of this heat is absorbed by greenhouse gases and reradiated back to the Earth's surface (Figure 2.1). This reradiated energy further warms the surface. The more greenhouse gases that are present, the more heat is absorbed and reradiated, and the hotter it gets.

The molecules responsible for this phenomenon are called greenhouse gases. The main naturally occurring and man-made greenhouse gases are carbon dioxide (CO_2), methane (CH_4), and nitrous oxide (N_2O). The most potent naturally occurring greenhouse gas is water vapor (H_2O). The amount of warming from water vapor is $\sim 60\%$ and from carbon dioxide it is $\sim 25\%$. Other greenhouse gases, including methane and nitrous oxide, account for $\sim 15\%$ (Kiehl and Trenberth, 1997). There are also a variety of man-made greenhouse gases that do not occur in nature. They are currently being emitted to the atmosphere, and some of these are absolutely horrendous greenhouse gases; in some cases over 9,000 times more potent than carbon dioxide.

Without the greenhouse effect the Earth would be largely uninhabitable. It is the greenhouse effect that makes complex life possible. The temperature of the Earth is determined by the balance of incoming radiant energy from the Sun and the emission of outgoing radiant energy from the Earth. If there were no oceans, life or greenhouse gases, then most incoming radiation would be re-emitted to space. Under these conditions the Earth's average global temperature would be $-18\,°C$ instead of the present $14\,°C$. The $32\,°C$ difference is due to the powerful greenhouse effect of water vapor that constitutes up to 4% of the atmosphere.

Other greenhouse gases are increasing, and these gases are causing the temperature to rise. Increasing the concentration of these gases increases the atmosphere's ability to block the escape of infrared radiation (heat). As the abundance of greenhouse gases increases, warming increases more and more. Too great a concentration of greenhouse gases can have dramatic effects on climate and significant consequences for life. Climates suitable for human existence do not exist above some minimum threshold level of greenhouse gas concentration. They exist within a limited range that makes complex life possible.

The greenhouse effect concept is not new. Almost 180 years ago in 1827 the French physicist Joseph Fourier conceived of the greenhouse effect, and in 1859 the Irish physicist John Tyndall invented the spectrophotometer to measure the absorption and reradiation of various wavelengths of light. He found that water vapor, carbon dioxide, and methane absorbed and re-emitted infrared radiation, but not nitrogen or oxygen (the main constituents of the atmosphere). In 1896 the Swedish chemist Svante Arrhenius theorized and calculated that a two-fold increase in atmospheric carbon dioxide could raise the air temperature about $5\,°C$ which is not far from today's super computer estimates. Since then scientists have periodically warned that increases of greenhouse gases could warm the Earth sufficiently to produce adverse consequences for humans and other species.

One of my favorites is the Englishman Guy Stewart Callendar who was not even a scientist, but a steam power engineer in the 1930s (Weart, 2003). His hobby was studying weather statistics, and he noticed that the numbers showed that the Earth was getting warmer. In 1938 he had the audacity to stand before the prestigious Royal Meteorological Society in London and tell them that the Earth was experiencing global warming. He also said that he knew why. Human industry was burning fossil fuels and releasing millions of tons of carbon dioxide that was trapping heat and changing the Earth's climate. How right he was! However, in those days his proposition reached deaf ears, and, of course, nothing came of it. There have been other farsighted individuals who have seen what was really happening to the Earth's climate, but were also ignored. To a certain extent this is still happening today. Unfortunately, that is the history of science. I salute those individuals who had the foresight to see the problem, but were ignored by their contemporaries.

11

THE CARBON CYCLE

The carbon cycle is the movement of carbon in many forms between the biosphere, the atmosphere, the oceans, and rocks. The cycle has various sinks where carbon is stored and different processes by which the various sinks exchange carbon. Without the carbon cycle, carbon dioxide and other carbon-bearing greenhouse gases would continuously accumulate in the atmosphere until our temperature reached values similar to those on Venus (480 °C). Carbon is continuously cycled between various reservoirs in the ocean, land, and the atmosphere (Figure 2.2). Most of this carbon comes from carbon dioxide. During the pre-industrial era the atmosphere contained about 560 billion metric tons of carbon (2,053 billion metric tons of CO_2 equivalent), but today it contains about 730 billion tons (2,677 billion tons of CO_2 equivalent) due primarily to human burning of fossil fuels and other human activities.

There is a strong and relatively rapid link (days to years) between the atmosphere, life, soils, and the upper ocean. However, the exchange between this rapid system and the deep ocean is much slower (several hundred years). The oceans absorb CO_2 at higher latitudes where the water temperature is cold, and release it near the tropics where it is warm. For each degree Centigrade rise in the ocean's surface temperature, there is a 4% decrease in the absorption of CO_2. Photosynthesis takes CO_2 from the

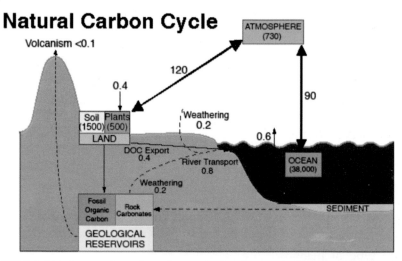

FIGURE 2.2 This diagram shows the exchange of carbon between the land, atmosphere, rocks and ocean (the numbers are in billion metric tons of carbon equivalent). Most carbon is stored in the oceans (~ 38,000 billion metric tons), and the exchange between plants and soil with the atmosphere is about 120 billion metric tons. See text for explanation. (Modified from the UN Intergovernmental Panel on Climate Change report, 2001)

atmosphere and transfers it to vegetation, but a process called *autotrophic transpiration* releases a smaller amount of CO_2 back into the atmosphere.

By far the largest reservoir of carbon is the deep ocean, which contains about 38,000 billion tons of carbon. In the upper ocean, the main form of carbon is dissolved carbon dioxide and small creatures such a phytoplankton. This accounts for another 1,000 billion tons of carbon. Atmospheric CO_2 dissolves in the ocean's surface waters where it undergoes rapid chemical reactions with the water. Only a small amount of CO_2 remains as carbon dioxide; the rest consists of various compounds of carbon. Both the CO_2 and chemical forms are known as *dissolved inorganic carbon*.

Dissolved inorganic carbon is transported by ocean currents called the *thermohaline circulation* (see Chapter 4). When it reaches high latitude regions the cold dense waters sink to the bottom of the ocean and eventually spread throughout the bottom of ocean basins. Hundreds of years later these waters return to the surface. Because more CO_2 can dissolve in cold water than in warm water, the cold dense water is rich in carbon. In turn, large quantities of carbon are moved from the surface to deep waters, a process called the *solubility pump*.

Dissolved inorganic carbon also is used by marine organisms. Phytoplankton takes up the carbon dioxide during growth and converts it to complex organic forms. Upon death these and other creatures transfer carbon from the surface to depth where it is eventually buried. This results in the very large accumulation of carbon in marine sediments ($\sim$38,000 billion tons).

On land, plants account for about 500 billion tons of carbon, and decaying organic matter accounts for another 1,500 billion tons. Photosynthesis is the process by which plants use atmospheric carbon dioxide. They convert CO_2 to compounds that make up the roots, leaves, and wood. However, some of this carbon is lost and becomes soil carbon when the plant dies or sheds its leaves. In the soil, microbes break down this carbon and release it back to the atmosphere in the form of CO_2; a process called *respiration*.

The carbon cycle is greatly affected by the distribution and types of vegetation. Different plant types store different amounts of carbon, but they favor different conditions and grow at different speeds. Per unit area, trees can store more carbon than grass, but they take much longer to grow. The local climate conditions, and how they change over time, determine which type of plant is more abundant at a certain location. Figure 2.3 shows the current absorption of CO_2 by continental regions. Desert regions obviously absorb the least, and forest and jungle regions the most.

Unfortunately, human activity has adverse affects on the carbon cycle by changing land use and, therefore, the amount of carbon that is stored by the biosphere. Forest clearing removes a large absorber of carbon dioxide and if the wood is burned or left to decay, then the carbon is released back to the atmosphere as carbon dioxide and methane. Disturbing vegetation such as deforestation can also lead to large amounts of carbon being lost from the soil which, in turn, impacts the fertility of the ground and affects future growth. Most of the human CO_2 released in the 19th century was from this type of activity, and it was not until about 1950 that fossil fuel emissions became the leading source of greenhouse gas emissions; however, these topics will be discussed in more detail in Chapter 7.

The largest reservoir of CO_2 is marine sediments and sedimentary rocks (primarily carbonates) that have accumulated over geologic time. This reservoir holds an estimated 10^{17} metric tons of carbon or the equivalent of about 3.67×10^{17} metric tons of CO_2. This is equivalent to a surface pressure of about 70 bars of CO_2, which is about 20% less than the current CO_2 atmospheric surface pressure on Venus (90 bars).[1] Today we are releasing the CO_2 stored in marine sediments back into the atmosphere by manufacturing cement from carbonate rocks.

In earlier centuries, the natural transfers of CO_2 were about 20 times greater than those due to human activity. The natural transfers at that time were nearly in balance; and the magnitude of the CO_2 sources closely matched the sinks. The atmospheric CO_2 content had been essentially constant for at least the last 10,000 years. However, human activity over the past 150 years has caused the atmospheric CO_2 content to rise by about 35%, and at the present time we are out of balance because of these greenhouse gas emissions by humans.

Climate change can profoundly change the carbon cycle. Although growth in CO_2 abundance increases photosynthesis, the accompanying rise in temperature increases plant and soil respiration rates, which reduces CO_2 storage. In some regions decreased rainfall brought about by climate change can reduce planet photosynthesis, thus decreasing the CO_2 uptake. There are numerous other examples of perturbations of the carbon cycle by global warming, and few of them are good. In later chapters we will discuss how certain carbon sinks are beginning to decline because of global warming.

WEATHER AND CLIMATE

People often confuse weather and climate; however, they are not the same. Weather is the *day-to-day* atmospheric conditions: snow, rain,

---

[1] 1 bar is the atmospheric surface pressure at sea level on Earth.

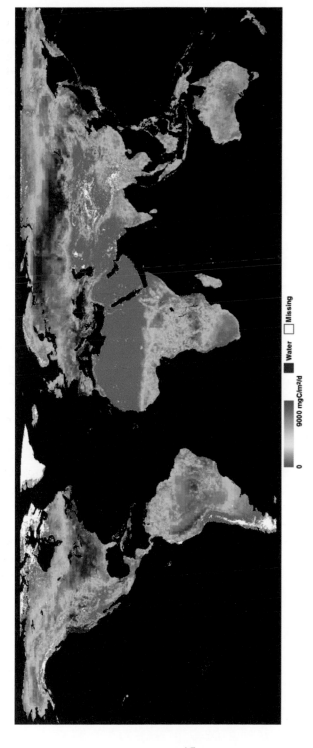

FIGURE 2.3 This map derived from satellite data shows the variation in absorption of carbon dioxide in continental areas. The brown areas absorb the least and generally correlate with desert or lightly vegetated regions, and the dark green areas absorb the most and generally correlate with heavily forested areas. (From National Oceanic and Atmospheric Administration)

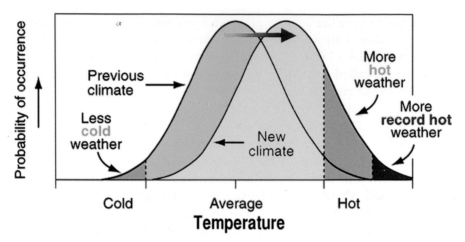

FIGURE 2.4. *This diagram shows a shift in climate from less cold weather to more hot and record hot weather. This is a climate shift similar to what is happening today. From the IPCC 4th Assessment Report 2007.*

tornadoes, hurricanes, or heat waves and cold spells. Climate is the *long-term average* atmospheric conditions, usually over a period of at least 30 years. Everybody knows that the climates of Miami, Florida and Stockholm, Sweden are very different. However, it rains in both Stockholm and Miami and the weather can be the same. Both cities can also have heat waves and cold snaps. This is weather, not climate. If winter temperatures in Miami and Stockholm become consistently warmer or colder than normal for relatively long periods of time, then this will be considered as a climate change. That is what global warming is all about: a consistent increase in the temperature with time, resulting in climate change. We can do nothing about the weather, but we are, in fact, changing the climate, and that, in turn, can intensify and increase the frequency of weather events. Figure 2.4 shows how a shift in climate causes changes in the frequency and intensity of temperature.

GLOBAL TEMPERATURE CHANGES

Small changes in the average global temperature result in large changes in global climate. We have established a standard baseline temperature against which present and past global average temperatures are compared. It is the climatological equivalent of sea level. The most frequently used baseline temperature is the average global temperature between 1951 and 1980;

this is a time when the average global temperature was fluctuating around moderate values compared to the past and present. This global average temperature is 14 °C (55 °F). Global average temperatures above or below this value are called *temperature anomalies*—positive or negative departures from this global average.

It does not take a large temperature anomaly to cause an enormous climate shift. We are accustomed to fairly large temperature changes between day and night, and in dry desert climates, for instance, the difference can be more than 22 °C between the daytime high and night-time low. Despite these large diurnal swings in temperature the global average is still about 14 °C. However, if the *global average temperature* dropped only 6 °C we would be in the depths of an Ice Age with ice sheets covering northern Europe, all of Canada, and the northern part of the United States. If the temperature rose by about the same amount we would be in a "Hot House," an abnormally hot period in the Earth's history with no ice and a much higher sea level.

NON-UNIFORM WARMING

People who are unfamiliar with climate change often believe that during a climate change the Earth cools down or heats up uniformly. This is not true. At present the Earth is heating up, but not uniformly. Some parts are warming faster than others. Most parts of the Northern Hemisphere at relatively high latitudes are warming faster than the Southern Hemisphere or tropics, and some areas in the north are warming considerably faster than others. A few areas at high southern latitudes are also warming rapidly. There is even one area that has cooled down: central Antarctica.

Climate has naturally occurring fluctuations that result in a saw-tooth temperature curve. Figure 2.5 is a diagrammatic representation of some possible climates. All have natural climate fluctuations, but some have long-term major climate changes. Our present increase in global temperature is not uniform: for a variety of reasons, some years will be cooler or hotter than others. Although the overall trend is for increased temperatures, there are some years that are warmer or cooler than adjacent years. For example, 1998 and 2005 were the hottest years on record since world temperature recordings began in 1880, but the years in between have been slightly cooler (2002 and 2003 were the second and third warmest years, respectively). The temperature trend, however, has been consistently upward (see Figure 6.5 in Chapter 6).

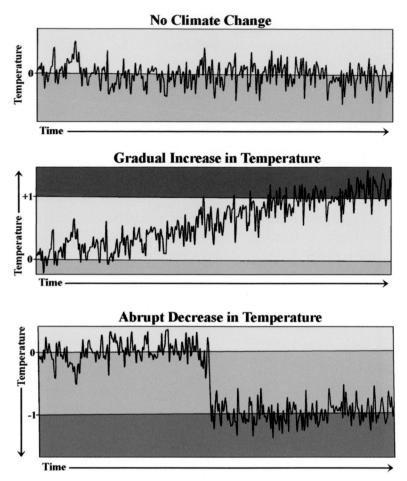

FIGURE 2.5 *This is a diagrammatic representation of two possible climate changes. In the top panel there is no climate change, only the natural variations shown by the serrated curve. In the middle panel there is a gradual increase in temperature along with natural variations. The bottom panel shows an abrupt climate change into cold conditions.*

FEEDBACKS

Feedback processes are extremely important to the magnitude and prediction of climate change. When one climate variable changes it can alter others in such a way that they change the processes responsible for the initial change. This circular response can reinforce the original stimulus, moving the whole system dramatically in one direction. For example, global warming can, and is, causing less snow cover in winter. This, in turn, leads to more sunlight being absorbed by the surface, causing even more warming, which causes less snow cover, and so on. This is called a *positive feedback* because it reinforces and increases global warming. On the other hand, warming could cause more evaporation and greater cloud cover that reflects more sunlight back to space, cooling the Earth until it reaches some equilibrium condition. This is called *negative feedback* because it tends to decrease warming or keep it the same. As we shall see, however, global warming is producing positive feedbacks in several important processes that far outweigh any negative feedbacks.

ABRUPT CLIMATE CHANGE

Some natural climate changes can be very rapid (Bard, 2002; National Research Council, Committee on Abrupt Climate Change, 2002). Before the more recent studies of ice cores it was thought that past climate changes took long periods of time to accomplish; centuries to millennia, or even millions of years. Although this is the case for many past changes, it has recently been discovered that some dramatic climate changes have occurred over a very short period of time—some in less than a decade. It is conceivable that the present rise in global temperature could trigger one of these abrupt climatic events, which is by far the worst consequence that could happen. Abrupt climate changes will be discussed in more detail in Chapter 5.

RADIATIVE FORCING

Radiative forcing is the quantity used to measure the amount of climate change. It is expressed in terms of watts per square meter (W/m^2). A positive value means that the atmosphere is heating, and a negative value that it is cooling.

If greenhouse gases increase, the lower atmosphere (the troposphere) warms as a result of absorbing and consequently emitting more infrared radiation (heat). Achieving a balance between incoming and outgoing radiation leads to the surface and troposphere warming, and the upper atmosphere (stratosphere) cooling, as shown in Figure 2.6. Radiative forcing is usually expressed in terms of the change in the average net radiation at the top of the troposphere, known as the tropopause. The troposphere extends from the surface to about 11 km. Within this layer the temperature gradually

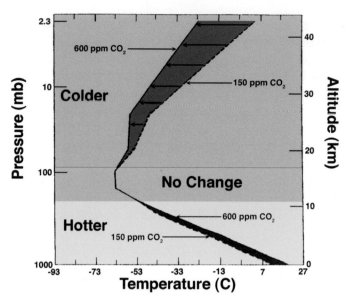

FIGURE 2.6 This diagram shows the change in temperature with height for two amounts of CO_2, assuming relative humidity and cloud cover remain the same. The lower atmosphere (troposphere) is warmed by the larger amount of CO_2, but in the stratosphere above a pressure of about 100 millibars (mb) there is cooling, and in a transition zone there is neither heating nor cooling. See text for explanation. (Modified from Fig. 20.4 in Trenberth, 1992)

decreases with height at an average rate of 6.5 °C per kilometer and, essentially, all important weather phenomena occur within this atmospheric layer. At present, radiative forcing is positive at about 2.0 W/m², and increasing. Although this may seem to be very little, the climate forcing between an ice age and today is only about −6.5 W/m², and we should remember that this amount of negative forcing resulted in a global temperature decrease of about 5 °C, which maintained the frigid temperatures of the ice age. It implies a climate sensitivity of about 0.75 °C for each W/m².

A small Christmas tree light is about 1 watt. To put the current radiative forcing in perspective, imagine that every square meter of the Earth's surface has two lit Christmas tree lights. Since the area of the Earth is 5.1×10^{14} m², the total amount of climate forcing is 1.02 billion megawatts—about the amount of electricity the entire world uses in 14 hours.

RADIATION BALANCE OF THE EARTH

The Earth's climate essentially depends on three basic parameters: (1) the amount of sunlight hitting the Earth; (2) the fraction of the sunlight reflected back to space; and (3) the amount of infrared radiation (heat) trapped by greenhouse gases. The least well-studied parameter is number (2).

Solar radiation interacts with Earth in several ways (Figure 2.7). About 20% is absorbed by the atmosphere and re-emitted to space; about 20% is reflected by clouds; and 9% is reflected by atmospheric dust and bright surfaces such as ice. The remaining 51% is absorbed by the surface, resulting in surface warming.

Much of this absorbed energy is eventually reradiated at longer infrared wavelengths as heat. About 17% of this heat escapes the atmosphere while the remaining 83% is absorbed by greenhouse gases (mostly water vapor) and reradiated back to the Earth's surface. This reradiated energy further warms the surface of the Earth. Currently the Earth has an energy imbalance; we are accumulating more energy than we are losing to space.

UNCERTAINTIES

There are many changes happening to Earth and the biosphere that are due to either natural variations in the climate, human impacts on the environment, or global warming. Deciding between these causes is not always easy, and in some cases change may be due to two causes working in concert. For example, coral reef degradation is known to be due to both human impacts and warming oceans due to global warming. Also, a single devastating storm system like hurricane Katrina cannot be attributed to global warming. Only if there is some systematic change in the long-term behavior of storms that corresponds to atmospheric or oceanic warming can the cause be attributed to global warming with any degree of confidence. One heat wave, a devastating flood, or one intense draught cannot be attributed to global warming unless there is some extraordinary circumstance that precludes some short-term natural variation. The 2003 European heat wave is an example of a possible global warming event because no such event has occurred in Europe for at least the last 500 years. Linking any one major event to a single cause is uncertain. In this book I have tried to avoid attributing any single event to global warming unless there is good reason to believe that it is at least partly responsible.

You should now have at least a rudimentary understanding of the fundamental aspects of climate change and how they interact. Before we go on to discuss climate change on Earth, it is important to put the subject in perspective by looking at climate conditions and changes on other planets, and why the Earth is such a unique and special place.

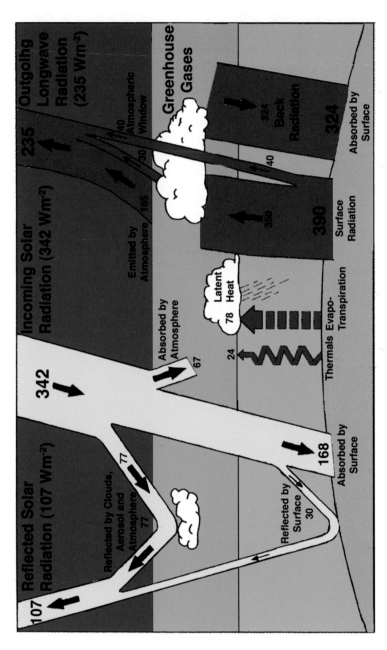

FIGURE 2.7 The energy balance is shown by the various processes that absorb and reflect heat. The yellow areas show the incoming solar radiation that is absorbed or reflected. The red areas are heat that is reradiated and lost to space by various processes or reradiated and absorbed by the Earth. The numbers are in watts per square meter. The total incoming solar radiation is 342 W/m^2. The net terrestrial radiation lost to space from the surface, clouds, and throughout the atmosphere balances out the total amount of absorbed incoming radiation. We are now out of balance due to our greenhouse gas emissions. See text for explanation. (Modified from the UN Intergovernmental Panel on Climate Change report, 2001, Fig. 1.3)

3

A PLANETARY
PERSPECTIVE

OVERVIEW OF THE SOLAR SYSTEM

ALL planets with atmospheres undergo weather changes of one kind
or another. The cloud patterns and atmospheric conditions on the
giant outer planets are constantly changing, and that is also true
for the inner planets (Figure 3.1). These changes reflect short-term
weather conditions, not climate changes. Before we examine the Earth's
changing climate it is useful to explore the climatic conditions on other
planets similar to the Earth in order to appreciate the uniqueness of Earth
and its climate system.

The Solar System is divided into two groups of major planets separated
by the asteroid belt: four outer or Jovian planets and four inner or
Terrestrial planets (Table 3.1). The Jovian planets (Jupiter, Saturn,
Uranus, and Neptune) are large Jupiter-like planets far from the Sun
consisting primarily of hydrogen and helium (low-density objects). The

FIGURE 3.1 *The solar system planets. The upper five are the terrestrial planets and the Moon, and the bottom four are the Jovian planets. Each group of planets is scaled to each other, but the groups have different scales. Three Earths would fit into Jupiter's Great Red Spot (the orange oval in the bottom half of the planet). (Courtesy of NASA)*

Jovian planets are from 5 to 30 AU distant from the Sun.[1] The Terrestrial planets (Mercury, Venus, Earth, and Mars), on the other hand, are

---

[1] Planetary distances are measured in Astronomical Units (AU). One AU is defined as the average distance between the Earth and Sun.

relatively small planets consisting of rock and iron (high–density objects). They are closer to the Sun, between 0.39 (Mercury) and 1.5 (Mars) AU.

TABLE 3.1 *Physical characteristics of the planets*

Planet	Mass (kg)	Radius (km)	Density (g/cm^3)	Distance from Sun (AU)	Period (years)
Mercury	3.3×10^{23}	2,439	5.43	0.39	0.24
Venus	4.8×10^{24}	6,051	5.20	0.72	0.61
Earth	5.9×10^{24}	6,378	5.52	1.00	1.00
Mars	6.4×10^{23}	3,396	3.93	1.52	1.88
Jupiter	1.9×10^{27}	71,492	1.33	5.20	11.8
Saturn	5.7×10^{26}	60,268	0.69	9.54	29.4
Uranus	8.7×10^{25}	25,559	1.32	19.2	83.7
Neptune	1.0×10^{26}	24,764	1.64	30.0	163.7

TABLE 3.2 *Major characteristics of the planets*

Property	Terrestrial planets	Jovian planets
Size	Small	Large
Density	Large	Small
Distance	Close to Sun	Far from Sun
Atmosphere	None to moderate	Very massive
Temperature	Warm	Cold
Composition	Mostly Rock and Iron	Mostly H and He
Number of satellites	3	Many (>70)
Composition	Rocky	Mostly icy
Rings	None	All

The major characteristics of the Jovian and Terrestrial planets are totally different (Table 3.2). Their various characteristics are the result of their origination in different parts of the Solar System. The Jovian planets, which formed in the cold outer reaches of the Solar System where volatile (easily vaporized) elements are very stable, consist primarily of the highly volatile compounds of hydrogen (H_2) and helium (He_2)—the main constituents of the Sun. In fact, they contain about the same proportion of hydrogen and helium as the Sun. The terrestrial planets formed in the hotter regions much closer to the Sun where volatile elements are rather unstable, but *refractory elements* (that is, they are difficult to vaporize) are stable. They, therefore, consist mostly of the refractory elements and compounds that make up rock and iron.

TERRESTRIAL PLANET ATMOSPHERES

We will examine only the terrestrial planets with atmospheres. Mercury and the Moon have no atmospheres because they are small objects with gravity fields too weak and dayside temperatures too high to hold any significant atmosphere. Only Venus, Earth, and Mars are large enough to have relatively dense atmospheres with weather and climate, and the atmospheric characteristics of these planets are listed in Table 3.3.

TABLE 3.3 Atmospheres of Venus, Mars, and Earth

Composition (%)	Venus[a]	Mars	Earth
N_2	3.5	1.6	78
O_2	–	0.13	21
Ar	–	2.7	0.93
CO_2	96.5	95.3	0.035
H_2O	–	~0.03	0–4
Surface pressure (bar)	90	0.008[b]	1
Surface temperature (°C)	480	−60 to −123[c]	60 to −130[d]
			Mean = 14

[a] Below 80 km; [b] average surface pressure; [c] average equatorial and polar diurnal temperatures; [d] temperature extremes

Venus

Venus is the classic example of a greenhouse planet. It has a dense atmosphere of CO_2 with clouds of sulfuric acid in its upper atmosphere (Figure 3.2). Venus is close to the Sun (0.72 AU), and its atmosphere is over 96% carbon dioxide. The atmosphere is so dense that its surface pressure is 90 times greater than the atmospheric surface pressure on Earth. This is equivalent to the pressure 915 meters (3,000 feet) under the ocean. Because of its proximity to the Sun and its dense greenhouse atmosphere, the surface temperature is an enormous 480 °C (896 °F)—hot enough to melt zinc. Furthermore, the temperature is isothermal, which means that it is the same everywhere on the planet. It is just as hot in the polar regions as at the equator. No water can exist on its surface or in its lower atmosphere.

We do not know if the climate was different earlier in Venusian history. Very early in its history Venus may have had a different atmosphere and a different climate, but certainly in the last billion years or so the climate has probably been the same as it is today. There is evidence that the planet has experienced multiple periods of global resurfacing, and the present surface

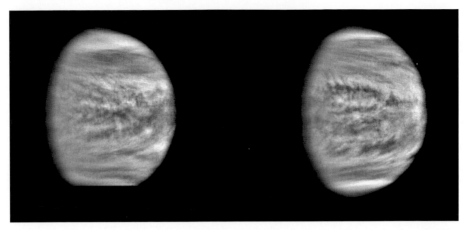

FIGURE 3.2 Venus atmospheric changes are shown in these images taken through the ultraviolet filter by the Galileo spacecraft on its way to Jupiter after a gravity assist from Venus. The images were taken about an Earth day apart and show changes in the cloud patterns. (Courtesy of NASA)

is largely a result of the last global resurfacing event. During these events there was an incredible amount of basaltic volcanism and tectonic activity. Most of the present surface consists of enormous basaltic lava flows that must have outgased huge quantities of CO_2. The last global resurfacing event probably took place some 500 million to a billion years ago. The planet probably acquired its present atmosphere from the massive outgasing of CO_2 during these resurfacing events. Today there is little internal activity compared to past resurfacing events.

On Earth, the second most abundant gas after water vapor in basaltic eruptions is CO_2. However, the CO_2 is largely absorbed by ocean water and life, and deposited as carbonate sedimentary rocks (limestone and dolomite). If all the CO_2 contained in carbonates were released, the Earth would have an atmosphere consisting of about 70 bars of CO_2. Without oceans and life on Venus there is no way for the CO_2 to be absorbed, and the planet continued to accumulate CO_2 to its present abundance and consequent high surface temperature. Today the climate is incredibly hot and dry everywhere on the planet. No significant climate changes have occurred in over a billion years.

Mars

Mars is a very different story. At present Mars has only a very thin atmosphere with an average surface pressure of 0.008 bar. This is equivalent to the pressure in the Earth's stratosphere at an altitude of 35.7 km (22.3 miles). The Martian atmosphere is 95% CO_2, and the water vapor content, although variable, is very low ($\sim 0.03\%$). Although the

27

atmosphere is largely CO_2, it has a negligible greenhouse effect because the atmosphere is so thin. Also, the planet receives much less solar irradiance than the Earth because it is farther away from the Sun. Consequently, Martian surface temperatures are very cold. The coldest temperature is $-148\,°C$ ($-193\,°F$) at the South Pole in winter, and the warmest is $21.8\,°C$ ($71\,°F$) at mid-latitudes in the summer. However, the average diurnal temperature is only $-60\,°C$ ($-76\,°F$) in equatorial regions and $-130\,°C$ ($-202\,°F$) in the polar regions (Table 3.3).

At present liquid water cannot exist on the Martian surface because of the temperature and pressure conditions. There are, however, clouds (Figure 3.3) of both frozen CO_2 and water ice crystals. Frozen CO_2 and water ice caps occur at the poles (Figure 3.4). The planet's rotation axis is tipped $25.2°$, almost the same as Earth's ($23.4°$). Therefore, Mars experiences seasons like Earth. However, they are almost twice as long as Earth seasons, because the Martian year is almost twice as long (Table 3.1). During the summer in the Northern Hemisphere the CO_2 ice cap sublimes off (goes directly from a solid to a gas without a liquid phase) leaving a residual water ice cap that is stable at summer polar temperatures. Much more water is stored deeper in the subsurface.

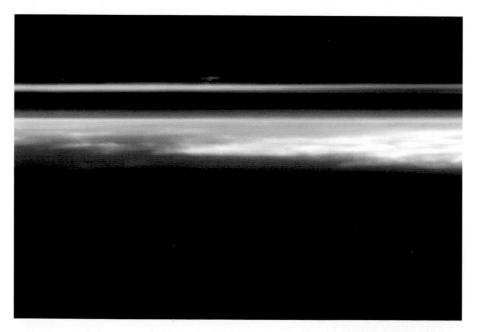

FIGURE 3.3 Cloud layers on the horizon of Mars imaged by the Mars Global Surveyor. Clouds on Mars consist of frozen carbon dioxide and water. (Courtesy of NASA)

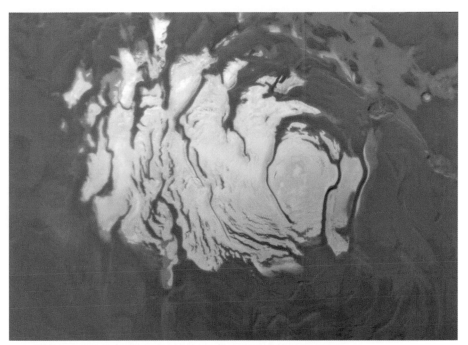

FIGURE 3.4 Mars' south polar cap viewed during southern summer. Most of the ice in this image consists of water. Most of the CO_2 ice has already sublimed off. (Viking image, courtesy of NASA)

Most fresh impact craters on Mars display ejecta blankets that are unique.[2] On the Moon, Mercury, and Venus, fresh craters have ejecta blankets that consist of a layer of continuous ejecta surrounded by strings and clusters of secondary impact craters. On Mars the ejecta blankets have lobate forms with radial flow lines (Figure 3.5). This type of ejecta blanket was formed by impact into subsurface water and/or ice. The water was incorporated into the ejecta and flowed like mud. Lobate ejecta craters are widely distributed on Mars and indicate that the subsurface has large quantities of water and ice over most of the planet.

There is convincing evidence that Mars has experienced short periods of very different climates than it has today, with thicker atmospheres of CO_2, warmer temperatures, standing bodies of water, and extensive ice sheets in the polar regions. Huge outflow channels tens of kilometers wide and hundreds of kilometers long are sites of the catastrophic release and flow of water (Figure 3.6). Almost all of these channels empty into the northern lowland plains of Mars. The source of this water is broken-up regions of the planet called *Chaotic Terrain*. This terrain was formed by the

---

[2]   An ejecta blanket is the material surrounding the crater and thrown out by the impact.

FIGURE 3.5 *The Martian impact crater Arandas with a lobate ejecta blanket formed by the flow of the ejecta. The ejecta flow was caused by water-rich ejecta from an impact into subsurface water and ice. The crater is 28 km in diameter. (Viking image, courtesy of NASA)*

catastrophic release of subsurface water that disrupted the surface. The discharge rate of the water can be estimated from the channel widths, depths, slopes, and other parameters. They are truly enormous. Some of the larger channels had peak discharge rates of 500 million to a billion

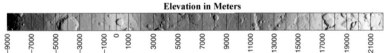

FIGURE 3.6 This topographic map of a portion of Mars shows large outflow channels caused by enormous catastrophic floods that issued from Chaotic Terrain and emptied into Chryse Planitia. The linear feature on the left is Valles Marineris, the largest canyon in the Solar System. The elevations are color coded so that blues and greens are low and yellows and reds are high. (Courtesy of NASA)

cubic meters per second. This is 25 to 50 times the peak discharge rate of one of the largest known terrestrial floods (Channeled Scablands, Missoula flood in Washington state). The amount of water disgorged by the Martian channels would be sufficient to fill large areas of Mars' northern lowland plains. The trigger that caused such large amounts of water to be

31

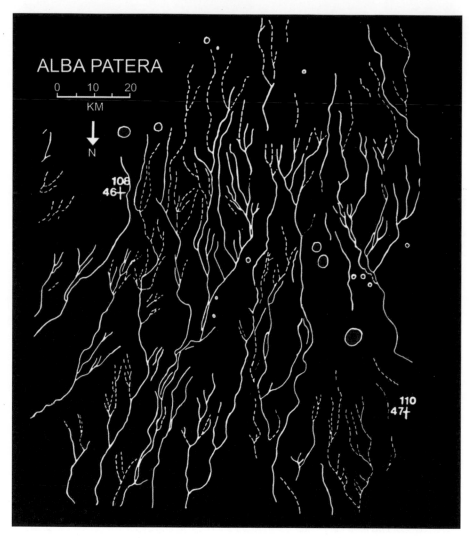

FIGURE 3.7 *A system of valley networks is shown on this map of a portion of the flank of Alba Patera volcano on Mars. They are the result of water runoff similar to stream channels on Earth. (Courtesy of Dr Virginia Gulick, NASA Ames Research Laboratory)*

disgorged from the interior is not well understood, but extensive volcanism is probably the most likely cause. Magma may have come into contact with ice and water infused with CO_2. This may have explosively released the water causing the catastrophic floods. The water probably accumulated in the northern plains to form a large ocean. The CO_2 released from the water may have been enough to cause a greenhouse atmosphere, warming the surface enough to make water stable. Age differences between the channels suggest that this occurred several times

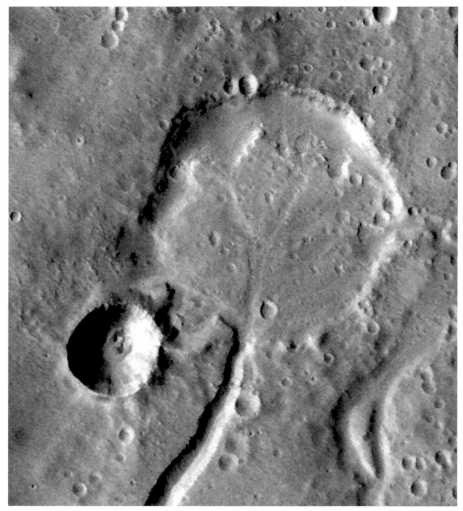

FIGURE 3.8 *This delta has formed in a crater 11 km in diameter. Several distributary channels can be seen on the delta. It was formed in water by the slow accumulation of sediment. The water has exited the right side of the crater and formed the channel on the right side of the image. (NASA Mars Odyssey Mission)*

in Martian history. The duration of these events may have only been a few thousands years, but they appear to have been periodic in nature.

In addition to the great outflow channels caused by catastrophic floods, there is a different type of channel called *valley networks* (Figure 3.7). This type of channel has branching tributaries and is much smaller than the outflow channels. They can only form by the relatively slow and steady flow of water over extended periods of time. Valley networks are concentrated in the heavily cratered uplands of the planet. Some of the networks have blunt, amphitheater-type heads, indicating that they were formed by a process called groundwater sapping where the flow of groundwater undermines the overlying rocks to cause them to collapse in their headwater reaches. Another type of channel is a system of complex branching tributaries that gradually taper away at their heads (Figure 3.7). These are very similar to runoff channels on Earth caused by the runoff of rain or melted snow. In certain craters there are deltas formed by river deposits in standing bodies of the water (Figure 3.8). Features such as these cannot possibly form under Mars' present climate. They strongly indicate that periodically the climate on Mars has been considerably milder than at present, with a denser atmosphere and warmer temperatures where water was stable. Recently the Mars Rover missions have discovered salts, sulfates and hydrated minerals that could only have been formed in water, thus corroborating the photogeologic evidence for past water on Mars.

Probably most astounding of all, was the discovery of very young channels and associated lava flows that, by their dearth of impact craters, suggests that they are very young, perhaps less that 10 million years old. This indicates that Mars is still geologically active, that the last water discharge event was relatively recent, and that such events may occur in the future. The Martian climate changes have been much more drastic than those on Earth, and their causes were very different.

Earth

Earth is unique in the Solar System (Figure 3.9). It is the only planet that has a permanent ocean and other standing bodies of water on its surface. The oceans cover about 70% of the surface. The water inventory on Earth is 98% oceans, 1.06% ice, 0.03% groundwater, 0.007% lakes and rivers, and 0.001% atmospheric water vapor. Water is a crucial ingredient for our climate. The oceans and atmospheric water vapor keep our planet habitable. The oceans are also the major sink for CO_2. The Earth's position between Venus and Mars is at a distance from the Sun where temperatures are mild and surface water is stable: the "comfort zone" of the Solar System.

Earth's atmosphere is also unique. It is the only atmosphere dominated by nitrogen (78%) and oxygen (21%). This atmosphere is a consequence of life that accounts for the oxygen content. Most of the atmosphere is concentrated in the troposphere, which reaches from the surface to an altitude of 11 km. This zone contains 75% of the atmosphere's mass, and all the weather and the most profound climate changes are in this region. The temperature decreases to $-55\,°C$ $(-67\,°F)$ at the top of troposphere, and it is heated by visible and infrared radiation. As we will see, the oxygen

FIGURE 3.9 *The Earth photographed by the Apollo 17 astronauts in 1972. Africa and the Arabian Peninsula are the continental landmasses. (Courtesy of NASA)*

and CO_2 abundances can vary over time and have an important influence on the climate.

Earth is the only planet that has plate tectonics. The outer rigid part of the Earth (the lithosphere) is fragmented into about 20 plates. They are moving slowly (2–10 cm/year) on a partially melted layer in the mantle called the *asthenosphere* (weak layer). The thickness of the plates varies from as little as 10 km in ocean basins to as much as 150 km under continents. New lithosphere is being formed at an interconnected global oceanic ridge system, and old lithosphere is being dragged down into the mantle and destroyed at subduction zones. The size, shape and location of individual plates are constantly changing. The changing continental positions and configurations have probably had a profound effect on past climates by changing the circulation of the oceans.

Earth is the only planet with a complex biosphere. Subsurface water on Mars and possibly a subsurface ocean on Europa may harbor primitive life forms, but only Earth has very complex life forms at its surface. This life has had a profound effect on the atmosphere, ocean, and land. It is constantly evolving, and is subject to naturally occurring events that have profoundly affected the way in which it evolves. Humans are now so abundant and polluting that they are affecting the Earth in ways that are upsetting the natural balance that has been established over millions of years.

The above characteristics of Earth also lead to a unique hydrological cycle, where evaporated ocean and lake water is recycled through the subsurface and surface of the continents, and the CO_2 emitted by volcanoes is absorbed by organisms and the oceans. It is completely different from the hypothesized cycle on Mars, and unique in the Solar System. As you will see, the Earth periodically goes through natural climate changes for a variety of reasons. However, for the first time in at least 500 million years, a single animal species is altering the atmosphere and causing a climate change.

4

TIME AND CAUSES

T HERE have been a number of significant climate changes in the geologic past. In order to appreciate what is happening to our climate now, it is important to understand past natural climate changes and their possible causes. We have just begun to delve into the realm of past climate changes in detail, but there is now a fairly good understanding of the major climatic conditions during the past 700 million years. Before that time we know little of the Earth's climate history. To appreciate the enormity of the time we are discussing it is important to understand the geologic time scale and to put it into some kind of perspective that we can grasp. Compared to geologic time the climate change we are experiencing now is occurring at lightening speed.

EARTH'S GEOLOGIC TIME SCALE

The Earth's history is divided into Eons, Eras, Periods, Epochs and various smaller divisions (Appendix A tells how rocks are dated on an absolute

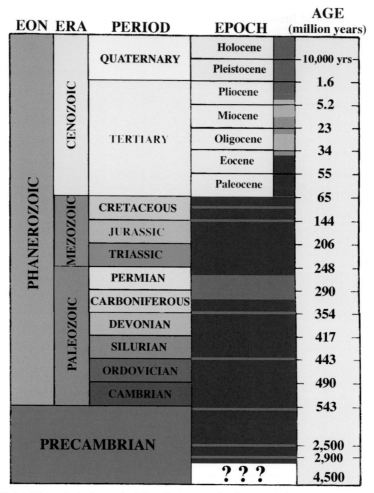

EON	ERA	PERIOD	EPOCH	AGE (million years)
PHANEROZOIC	CENOZOIC	QUATERNARY	Holocene	10,000 yrs
			Pleistocene	
		TERTIARY	Pliocene	1.6
			Miocene	5.2
			Oligocene	23
			Eocene	34
			Paleocene	55
	MEZOZOIC	CRETACEOUS		65
		JURASSIC		144
		TRIASSIC		206
	PALEOZOIC	PERMIAN		248
		CARBONIFEROUS		290
		DEVONIAN		354
		SILURIAN		417
		ORDOVICIAN		443
		CAMBRIAN		490
				543
PRECAMBRIAN				2,500
				2,900
			? ? ?	4,500

FIGURE 4.1 *The geologic time scale. The red areas are "hot houses" and the dark blue areas are "ice houses" as discussed in Chapter 5. In the Tertiary, the orange, light green, and light blue areas are progressively cooler periods. The thicknesses of the time periods are not proportional to their time spans. (Modified from Van Andel, 1994, Fig. 2.4)*

time scale). Figure 4.1 is an abbreviated version of the geologic time scale. It also shows the cool (Ice House) and warm (Hot House) periods in Earth's history in blue and red, respectively. The least understood part of Earth's history (the Precambrian Eon) comprises 88% of geologic time. The Phanerozic Eon is divided into three Eras, the Paleozoic (6%), the Mesozoic (4%), and the Cenozoic (1.4%). We know much less about the Precambrian because it has been largely reconstituted by plate tectonics. All of the Earth's ocean basin lithosphere (60% of the surface) has been formed by seafloor spreading during the last 200 million years (the last 4% of Earth's history).

Consequently, the history of 60% of Earth prior to 200 million years ago has been completely lost. Large areas of the continental regions have been melted many times by processes associated with subduction, crustal deformation, and impact. However, the continental Precambrian shields yield the oldest known rocks on Earth, and are the only source of knowledge about our early history. However, rocks older than 3.5 billion years are extremely rare. The oldest known rock is a 4-billion-year-old NW Canadian metamorphic rock called *gneiss*. The oldest known fossil life are bacterial organisms 3.45 and 3.46 billion years old in Western Australian rocks.

Our climate information is mostly confined to the last 700 million years. Obviously the closer we are to the present the better our knowledge of Earth's climate. Our best information on climate change is for the latest two Epochs comprising the Quaternary Period of the Cenozoic Era. These are the Pleistocene Epoch (1.6 million to 10,000 years ago) consisting of the most recent ice ages, and the Holocene Epoch that represents the present interglacial period (10,000 years ago to present).

Our life span averages only about 75 years, so it is very difficult to grasp the magnitude of 1,000 years, much less a million or a billion years. One way to appreciate the enormous span of geologic time and the rapidity of our current climate change is to compress it into a span of only one year (Table 4.1).

TABLE 4.1 The geologic time scale relative to one year

Eon or Era	Absolute time		Relative to a year	
	Duration (million years)	Percent geologic time	Starting time	Duration
Precambrian	3,996	88.0	January 1	10 mo. 17 d
Paleozoic	296	6.5	November 18	24 days
Mesozoic	183	4.1	December 12	15 days
Cenozoic	65	1.4	December 26	5 days

On this time scale one day equals about 12.4 million years, one hour about 518,000 years, one minute 8,638 years, and one second 144 years. The Precambrian starts the morning of January 1 and ends November 17. In other words, the least known part of Earth's history lasts over 10.5 months. It covers most of winter, all of spring, all of summer, and almost half of autumn. The Paleozoic begins November 18 and ends December 11, the Mesozoic ends 15 days later on December 26, and the Cenozoic ends 5 days later at midnight on December 31. All of human existence is squeezed into the last 46 minutes of December 31.

On this time scale the beginning of human evolution, which began about 7 million years ago, occurs on the last day of the year (Table 4.2). In fact, it starts at 10:30 a.m. December 31. The earliest humans (*Homo* species) develop about 7:34 p.m., and advanced human evolution (Early *Homo sapiens*) begins at 11:14 p.m. Modern humans (*Homo sapiens sapiens*) develop at 11:46 p.m. Civilization (agriculture) begins about one minute before midnight and the Industrial Revolution begins 1.7 seconds before midnight. On this time scale, your life span is about 0.5 second long, and the current (since about 1980) rapid rise in temperature began just 0.1 second ago.

TABLE 4.2 *Major human events on a relative time scale of one year*

Major event	Years before present	December 31	
		Starting time	Duration[a]
Human development begins	~7 million	10:30 a.m.	13 hr, 30 min.
Homo species begins	2.3 million	7:34 p.m.	4 hr, 26 min
Early *Homo sapiens*	400,000	11:14 p.m.	46 min
Homo sapiens sapiens (us)	125,000	11:46 p.m.	14 min
Civilization begins (Agriculture)	12,000	11:58:21 p.m.	1 min, 21 sec
Bronze Age begins	5000	11:59:34 p.m.	35 sec
Early writing	4000	11:59:32 p.m.	28 sec
Industrial Revolution (technology)	250	11:59:58.3 p.m.	1.7 sec
Human lifetime	75	11:59:59.4 p.m.	0.6 sec
Current rapid warming	25	11:59:59.9 p.m.	0.1 sec

[a] Starting time to the present.

NATURAL CAUSES OF CLIMATE CHANGE

The number of natural processes that can change the climate, can be roughly grouped into long-term and short-term causes. Long-term causes take thousands to millions of years, but short-term causes take only days to less than a century. In some cases two or more phenomena work in concert to cause a climate change. Appendix B describes the methods used to determine past climates.

Abundance of Greenhouse Gases

The amount of greenhouse gases in the atmosphere is probably the most obvious cause of climate change. The most important greenhouse gas is carbon dioxide (CO_2) because it is the most abundant. However, methane (CH_4) may have played a major role in some climate changes. There can be natural increases or decreases in these gases depending on the degree of biological activity and other factors. Figure 4.2 shows the estimated variation of oxygen (O_2) and CO_2 over the Phanerozoic Eon. The scale of these changes is still uncertain but their existence is well established. Although the abundance of O_2 is extremely important to evolution and a good indicator of biological activity, it is not a greenhouse gas and, therefore, of no consequence to climate change. The balance between CO_2 and O_2 is the result of the biologic cycle, which depends on the interaction of these gases with life and its waste products. Oxygen is produced by the photosynthesis of plants. It is taken up by animals, oxidized and emitted as carbon dioxide. The CO_2 is then taken up by plant photosynthesis to complete the process.

If biological productivity is low, then CO_2 is not taken up as efficiently and the CO_2 content of the atmosphere increases while the O_2 abundance decreases. Conversely, if the productivity is high, then CO_2 declines and O_2 increases. For example, the huge production and deposition of organic matter during the Carboniferous Period (the name derives from the huge carbon deposits) probably led to the large increase of O_2 and corresponding decrease of CO_2 (see Figure 4.2).

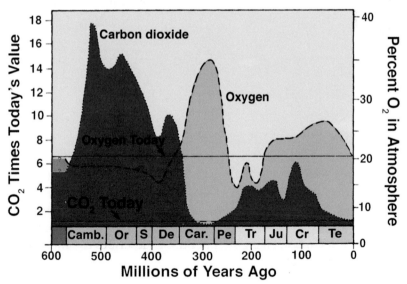

FIGURE 4.2 *This diagram shows the oxygen (blue) and CO_2 (red) contents over geologic time compared to today's values. There are fairly large uncertainties of the absolute values, but the relative abundances are well established. High oxygen levels indicate high biological activity, and high CO_2 levels indicate high temperatures. (Modified from Van Andel, 1994, Fig. 14.6)*

41

Two other processes linked to plate tectonics can also change the level of CO_2 in the atmosphere over tens of millions of years. One is the atmospheric input from volcanic eruptions and the other is the weathering of silicate rocks. Beside water vapor, the main gas emitted by volcanic eruptions is carbon dioxide. Variations in the intensity and frequency of volcanism with time will change the CO_2 level. Variations in the extent of silicate rock exposure over time can also affect the abundance of atmospheric carbon dioxide. Large areas of exposure will deplete the abundance more than small areas. The combined effects of biological productivity, volcanic activity and silicate rock weathering can change the various proportions of CO_2 and O_2 in the atmosphere.

There is also a strong possibility that methane (CH_4) locked up in ocean sediments has been released catastrophically to raise the temperature to high levels in short periods of time. For example, stable carbon isotope ratios ($^{13}C/^{12}C$) from ocean sediment cores indicate that about 1,200 to 2,000 billion metric tons of methane were released into the atmosphere about 55 million years ago. The gas was released from methane-rich ocean sediments probably by global warming that pushed the ocean–atmosphere system past a critical threshold. This powerful greenhouse gas probably caused the extreme warming at the end of the Paleocene Epoch (see Chapter 5).

Increases or decreases in greenhouse gases can thus be long term or very short term. Currently the increase in greenhouse gases is happening very rapidly, even on a human time scale. There is also a possibility that the current rapid increase in temperature could trigger a rapid release of methane and a devastating rise in temperature.

Major Volcanic Eruptions

Very large volcanic eruptions can change the climate in a very short period of time. The injection of large amounts of volcanic dust into the atmosphere produces a global layer of dust that reflects sunlight and cools the Earth. There have been a number of large historic eruptions that have cooled the temperature, but in order for such eruptions to cool the Earth on a global scale they must occur in the equatorial regions to ensure that the dust is distributed uniformly around the Earth.

The most recent large volcanic event that cooled the Earth was the 1991 eruption of Mt Pinatubo in the Philippines. It injected so much ash into the atmosphere that the tropospheric temperature was lowered by about 0.2 °C for about two years (see Figure 6.5(a)). Most of this cooling was over continental areas. When the ash settled down in about two years, temperatures returned to their pre-eruption values. Other major eruptions that probably changed history are discussed in Chapter 6.

Large Asteroid or Comet Impacts

Just as major volcanic eruptions, large impacts can eject enormous quantities of dust into the atmosphere and cool it down. As impacts take place in a matter of seconds to minutes, the effect on the atmosphere is almost instantaneous. If the impact is large enough a portion of the ejecta leaves the atmosphere and returns at relatively high velocities heating the atmosphere sufficiently to cause massive fires on almost a global scale. The atmospheric dust reflects much of the Sun's radiation to drastically cool the Earth to near freezing temperatures. The cooling can last several years, depending on the size of the impact. It can also eject enough CO_2 into the atmosphere to cause global warming after the dust settles if the impact is in carbonate rocks such as limestone.

Very large impacts have caused cataclysmic climate changes that resulted in mass extinctions of animal and plant life.[1] The degree of cooling is probably determined by the size of the impact: small impacts that do not eject much dust into the atmosphere would cause much less cooling for a shorter period of time.

Changes in the Thermohaline Circulation of the Oceans

The oceans of the world cover 71% of the surface and affect the climate in very important ways. Most of the Sun's radiation on the Earth's surface goes into the oceans to heat their surface waters. Because of their large heat capacity and circulation, the oceans store enormous quantities of heat and redistribute it before it is released to the atmosphere.

The distribution of the oceans' heat is carried out by the *Thermohaline Circulation*. The terms *thermo* and *haline* mean "heat" and "salt". Most people know that the ocean temperatures are greater in the tropics than at high latitudes, but they usually do not realize that the salt content also varies from region to region. For example, at mid-latitudes in the Atlantic between the Gulf of Mexico and West Africa the salinity is greater than at higher latitudes. The reason is that the Sun warms the equatorial regions more than at higher latitudes enhancing the oceans' evaporation and making tropical ocean water saltier. Figure 4.3 shows a simplified diagram of the oceans' thermohaline circulation. This large-scale circulation is due to a combination of (1) wind-driven currents in the upper few hundred meters of the sea, (2) currents driven by fluxes of heat and fresh water to the surface with subsequent interior mixing by heat and salt, and (3) tides

---

[1] The impact responsible for the Cretaceous/Tertiary extinction of the dinosaurs is discussed in the section on Mass Extinctions in Chapter 5.

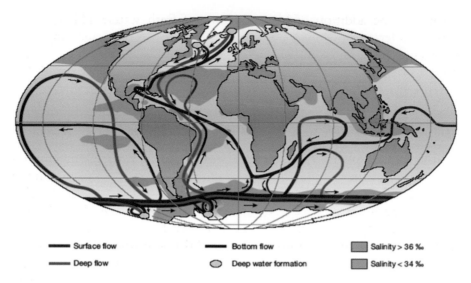

FIGURE 4.3 This diagram of the thermohaline circulation shows the warm surface currents in red, and the cold deep and bottom currents in blue and brown. The orange areas are where warm currents are sinking to form deeper circulation. The variations in ocean salinity are also shown. (See text for explanations.) From S. Rahmstorf, 2000.

caused by the gravitational pull of the Moon and Sun. In the diagram, the near-surface waters (red lines) flow to deep-water formation regions. The North Atlantic region is one area where the heat is released to the atmosphere. Here the system recirculates to deep or bottom currents (blue and brown lines).

The circulation carries heat to high latitudes to warm the atmosphere. The climate in the North American Atlantic region, which includes Europe and eastern North America, is controlled by currents that carry heat northward from the tropics. This conveyor belt is caused by saltier (therefore heavier) surface water that sinks to great depths off Greenland. This heavier water is replaced by warm water from the tropics that, in turn, heats the North Atlantic region. Without this warming, the relatively mild climate of northern Europe might be as cold as northern Canada.

There are at least two ways of changing the thermohaline circulation to drastically alter the climate. One way is to alter the configuration of the continents by continental drift and tectonic activity. For example, the closing of the Isthmus of Panama about 3 million years ago probably changed the circulation and aided the drastic cooling that followed.

A second way to change the circulation is by altering the salinity of the oceans by the addition of large amounts of fresh water. This causes a density difference (salt water is heavier than fresh water) allowing large changes in heat transport to occur. For example, large amounts of fresh water from melting glaciers introduced into the North Atlantic could shut down the transport of heat to high latitudes. Computer simulations of such an event under past atmospheric and insolation conditions result in a lowering of surface temperatures by as much as 6 °C in the Northern Hemisphere, plunging it into an ice age. Ironically, at the same time the Southern Hemisphere warms slightly because the sea surface temperatures increase. The time scale for changes in the thermohaline circulation are in the order of thousands to millions of years, unless there is a catastrophic release of fresh water into the North Atlantic. For example, about 17,500 years ago during the coldest interval of the last Ice Age large numbers of icebergs were discharged into the North Atlantic causing the thermohaline circulation to collapse and the regional temperature to plummet (McManus et al., 2004). This regional climatic extreme began suddenly and lasted for 2,000 years. Another slowdown of overturning circulation caused a cold period 12,700 years ago, lasting more than 1,000 years. These two cold periods were followed by a rapid acceleration of the circulation, resulting in significantly warmer climates over northern Europe and the North Atlantic region.

Continental Drift

The positions and configurations of the continents have not always been the same as they are today. Plate tectonics resulted in the break-up, joining and movement of continental masses over time, and this has had important effects on the Earth's climate. Figure 4.4 shows the evolution of the continent–ocean distribution over the past 500 million years. These distributions are derived from reconstructing the magnetic pole positions over time from magnetized minerals in rocks of various ages. In this way so-called "polar wandering" paths can be constructed for the various continents. In fact, it is not the magnetic poles that have wandered, it is the continents that have moved with respect to the magnetic poles. When climatic information is derived from ancient sediments, we need to know where these sediments were deposited in order to put the Earth's climate history into context. For example, we will see that extensive glacial deposits were laid down in the equatorial regions about 550 to 700 million years ago, although those rocks are not now near the equator.

As already mentioned, continental drift and the closing or opening of straits like Panama can alter the thermohaline circulation to change climates. Furthermore, when continental masses are grouped together into

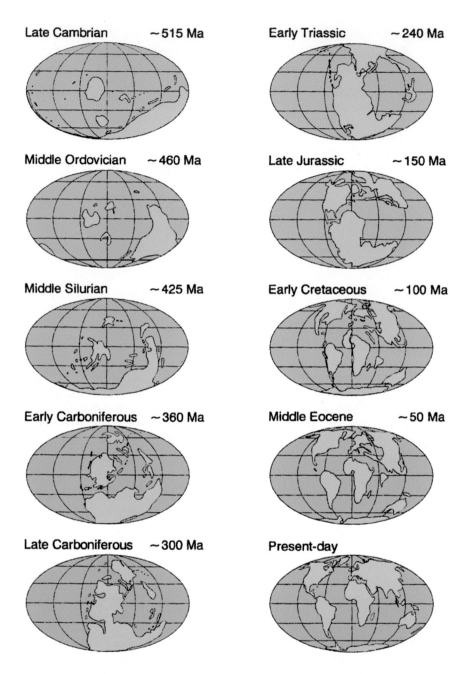

Late Cambrian ~515 Ma

Middle Ordovician ~460 Ma

Middle Silurian ~425 Ma

Early Carboniferous ~360 Ma

Late Carboniferous ~300 Ma

Early Triassic ~240 Ma

Late Jurassic ~150 Ma

Early Cretaceous ~100 Ma

Middle Eocene ~50 Ma

Present-day

FIGURE 4.4 These maps show the position of continental land masses from the Late Cambrian to the present. The position and size of the continental masses, in addition to other factors, control the ocean's circulation and the climate. They change on a time scale of millions of years. (Modified from Scotese, 1997)

46

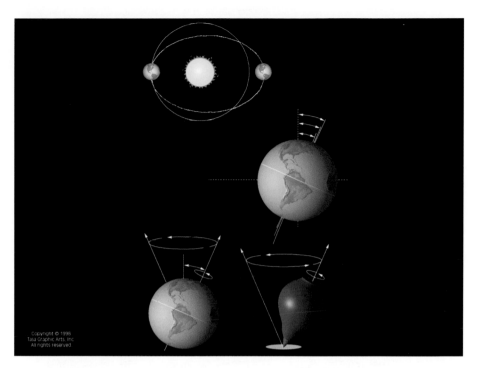

FIGURE 4.5 *The changes in the Earth's orbit and rotation are shown in this diagram. Top diagram shows how the eccentricity of the Earth's orbit can change (scale is highly exaggerated), The middle diagram shows variations in the tilt of the rotation axis (obliquity), and the bottom diagram show changes in the precession of the rotation axis similar to a spinning top. All of these motions affect the Earth's climate on time scales of thousands of years. (Courtesy Tasa Graphic Arts, Inc. with modifications)*

a single continent or a congregation of continents they can provide a locus for glaciation in the interior regions of the continental mass. The time scale of these changes are in the order of millions of years.

Changes in the Earth's Motions

The Earth's motions are not constant. They vary over a 20,000 to 100,000 year time scale and determine the amount of heat the Earth receives from the Sun (the amount of *insolation*). These motions are shown diagrammatically in Figure 4.5 and include the shape of the Earth's orbit around the Sun (called the *eccentricity*), the tilt of the Earth's rotation axis (called the *obliquity*), the time of *perihelion* (that is, when the Earth is closest to the Sun), and the *precession* of the Earth's rotation axis (the circular movement of the axis with respect to the Earth's orbital plane). Figure 4.6 is a diagram of (a) the changes in eccentricity, (b) obliquity, and (c) time of perihelion over time. The

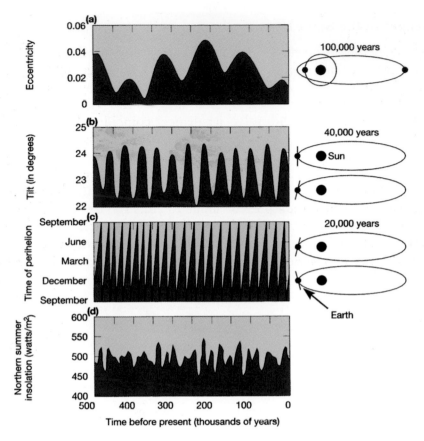

FIGURE 4.6 *This diagram shows the variation in Earth's orbital parameters and the variation of its axial tilt illustrated in Figure 4.5. These variations are known as the Milankovitch parameters that influence the onset and termination of ice ages. The eccentricity (a) is the departure of the Earth's orbit from a circle. Times of maximum eccentricity are roughly 100,000 years apart. The variations of the tilt of the Earth's axis is shown in (b). The time of perihelion (c) involves the Earth's axial tilt at its closest approach to the Sun. The amount of sunlight at 60 to 70 °N. latitude during the summer (d) is shown for variations of the Earth's orbital parameters. (After Covey, 1984)*

change in the insolation is also shown in the figure. These changes have been invoked to account for the periodicity of the Pleistocene ice ages. Although they have probably contributed an effect, it is unlikely that they are the sole cause of the periodicity. The amount of insolation change by itself is inadequate to initiate major cold or hot periods in Earth's climate history. This will be discussed in more detail in the section on the Pleistocene ice ages (see page 63).

Changes in the Sun's Irradiance

The energy output of the Sun (*irradiance*) is not constant. During the last 1,800 years the Sun has gone through nine cycles of changes in brightness or energy output. These long-term variations account for only about 0.1% of the total irradiance, but there is some evidence that they have affected the Earth's climate. The solar irradiance coincides with sunspot maximums and minimums. During minimums the irradiance is less and during maximums it is more. There are two sunspot cycles: the short-term cycle is 11.2 years, and the longer cycle is about 22 years. Part of the so-called "Little Ice Age" occurred during the Maunder Minimum when there were virtually no sunspots for about 70 years and the average global temperature was about 0.5 °C less than in 1970. However, the Little Ice Age lasted about 500 years from 1350 to 1850 AD. We are now in the Modern Maximum where the solar irradiance is at a maximum. Figure 4.7

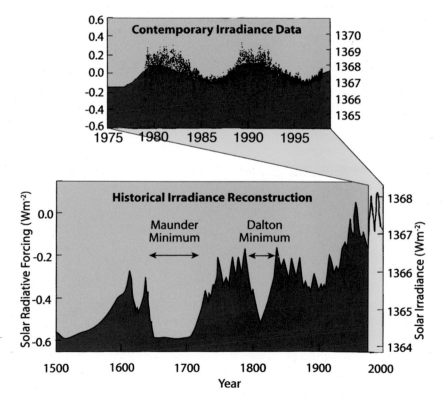

FIGURE 4.7 *This diagram shows the variation in solar irradiance from 1500 to 2000. Some of the minimums correspond to parts of the cold periods on certain parts of the Earth. Today's high solar irradiance can only account for a fraction of the current temperature rise. (Modified from Lean et al., 1995, and Rind et al., 1999)*

shows the changes in solar irradiance from 1500 to 2000. The present high solar irradiance cannot be the sole cause of the current global warming. It is about 10 times too small to account for the observed warming. On a geologic time scale these changes are short, but on a human time scale they are intermediate in length.

SUMMARY

This may seem like a bewildering array of processes that cause climate changes, and that almost any one or some combination of these processes could be the result of the present global warming. However, only two processes operating today cause warming fast enough to result in the present global warming: (1) increase in greenhouse gas abundance, and (2) increase in solar irradiance. While higher solar irradiance increases the global temperature, it is not strong enough to cause the entire rapid temperature increase experienced at present. The rapid increase in greenhouse gases is almost certainly the main cause of current global warming.

5

A HOT AND COLD PAST

THE Earth's climate can be divided into two general conditions, Hot Houses and Ice Houses. These general conditions are also shown on the geologic time scale in Figure 4.1. Figures 5.1 and 5.2 are diagrammatic representations of the distribution of these general climates over geologic time. The red areas are Hot Houses, and the blue areas are Ice Houses. We are currently in an interglacial period of an Ice House. Hot Houses are periods when the average global temperatures are significantly hotter than present, and Ice Houses are very cold times, or alternating glacial and interglacial periods. Interglacial periods are still considerably colder than a Hot House. In a Hot House there are no glaciers or ice sheets on Earth, and Antarctica would be covered with pine forests as it was in the Cretaceous Period when dinosaurs roamed that region of the world. Snow would be very rare, only occurring at the very highest latitudes or very highest altitudes. Figures 5.1 and 5.2 are only representations of prolonged dominant climatic conditions. In fact there

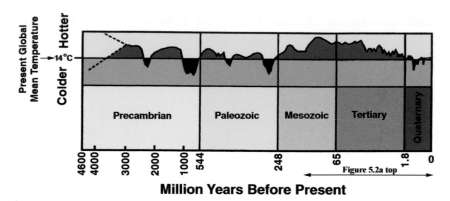

Million Years Before Present

FIGURE 5.1 *This diagram shows the general variations in the Earth's temperature throughout geologic time referenced to today's global average temperature. Red areas are hotter than today and blue areas are somewhat similar to today's Ice House with alternating glacial and interglacial periods. Most of Earth's history has been spent in much warmer conditions called "hot houses". The double arrow indicates the time span at the top of Figure 5.2a. (Modified from Cook et al., 1991)*

were limited cold periods in some Hot Houses and limited warm periods in some Ice Houses. The diagrams are meant to show only the dominant climatic regime in Earth's later history. That history has been dominated by Hot Houses with temperatures well above those today and little or no ice on Earth. Ice Houses have apparently been relatively rare throughout most of geologic history.

Precambrian Climates

The earliest occurrence of sedimentary rocks is about 3.8 billion years ago (the Isua metasedimentary rocks in west Greenland). At that time it is believed that the climate may have been warmer than today. Zircons found in sedimentary rocks are from earlier periods and range from 4.0 to 4.4 billion years old. Oxygen isotope ratios in these zircons are high, indicating that they formed in the presence of water (Valley et al., 2002). This suggests that oceans were present immediately after the formation of the Earth and that the climate was relatively mild. Planetary studies show that a cataclysmic bombardment of asteroids occurred from about 3.8 to 3.9 billion years ago (Strom et al., 2005). The cause of that bombardment was probably the migration of Jupiter and Saturn to their present positions, causing unstable regions to sweep through the asteroid belt ejecting them into the inner Solar System (Gomes et al., 2005). The large basin-forming impacts probably vaporized any oceans, and probably caused horrendous climatic conditions (first extreme heat and then extreme cold). After the bombardment ended, oceans returned and so did a relatively mild climate.

The first relatively well-known climatic period from the geologic record was a glacial period that occurred about 2.9 billion years ago. For at least part of the time between 2.2 and 2.4 billion years ago there was also widespread glacial conditions. From the presence of glacial deposits, it is believed that there were three discrete glacial epochs within this period.

The best-documented glaciations in the Precambrian occurred between 800 and 550 million years ago. At that time there was a large supercontinent in the equatorial regions. Two phases of glaciation with low latitude deposits (equatorward of $10°$) indicate that much of the Earth was frozen. One phase occurred at about 750 million years ago, and the other between about 620 and 550 million years ago. These periods of near global glaciation have been collectively termed the "Snowball Earth" (Kirschvink, 1992; Hoffman et al., 1998). At that time the Sun's irradiance was only about 95% of the present, and a large part of Earth may have been ice covered and reflecting much of the solar radiation. In this case it would be very difficult to ever get out of this frozen condition. In fact, carbon isotopic data indicate a collapse of oceanic photosynthesis for millions of years as a result of this extensive glaciation. Because of the lack of photosynthesis to take up CO_2, volcanic eruptions would lead to an accumulation of CO_2 large enough to produce a strong greenhouse effect, melt the ice, and possibly reach extreme greenhouse levels equivalent to a Hot House (Crowley et al., 2001). Recent studies of sedimentary rocks covering this glacial period indicate that it actually consisted of alternating warm and cold periods rather than just one long near-global glaciation.

The initial cause of this extensive glacial and interglacial period is not well understood. Possibly the lower solar irradiance coupled with feedback cooling from the reflection of solar radiation by bright ice might result in a near global glaciation. Another suggestion is that extensive glaciation was due to the location of continental masses in the tropical regions during a high obliquity (tilt) of the Earth's rotation axis ($>54°$) (Donnadieu et al., 2002). This would have been aided by the lower solar irradiance at that time. In any event, this extensive frozen stage appears to be confined to this period in Earth's history. The emergence of Earth from the first frozen period 700 million years ago may have resulted in a rapid diversification of higher plants and animals with soft bodies. The emergence from the second period at the end of the Precambrian may have led to the Cambrian explosion of complex life forms involving shells and skeletons with all the basic characteristics of more recent animal life (Lenton and Watson, 2004).

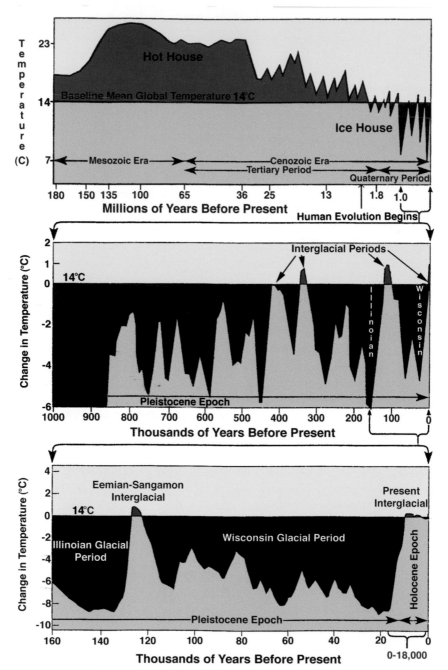

FIGURE 5.2(a) Changes in the Earth's temperature through time. Red areas are warmer and dark blue areas are colder. The top diagram spans the period from about the lower Mesozoic Era to the present. The two bottom diagrams are successive enlargements of the present Ice House during the Pleistocene and Holocene Epochs. (Modified from Cook et al., 1991)

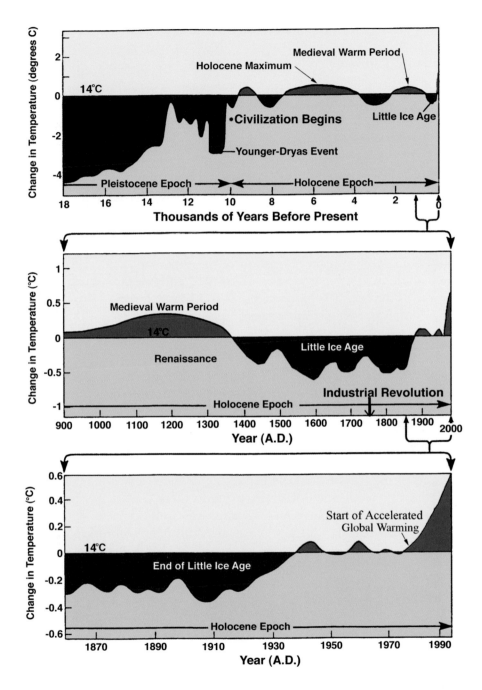

Figure 5.2(b) The top diagram shows the end of the last ice age and the present interglacial period (Holocene). Each of the enlargements show successively more detail, including current global warming. (Modified from Cook et al., 1991)

55

PHANEROZOIC CLIMATES

Paleozoic Climates

For the first 300 million years of the Paleozoic Era, the climate was significantly warmer than today. Near the end of the Ordovician and the beginning of the Silurian there was a brief glacial period. This evidence is found in rocks of those ages containing glacial deposits now in the Sahara desert, but which were close to the South Pole at the time they were deposited. There was also a glacial period during the Late Devonian/Early Carboniferous that lasted from about 360 to 350 million years ago. The Carboniferous saw a precipitous drop in temperature that formed the Permo–Carboniferous glaciation from about 330 to 250 million years ago. This period was one of the colder periods in Earth's history. The later stages coincided with the formation of the supercontinent called Pangea when all the Earth's continents were merged into one stretching from the equator to the South Pole. Large portions of Pangea were covered by ice, but other areas were probably significantly warmer.

Mesozoic Climates

Figures 5.2(a) and 5.2(b) are diagrammatic representations of the climatic conditions at progressively higher resolutions from the Jurassic Period in the Mesozoic Era to the present. In these diagrams the general best estimates of the temperature anomalies are given with respect to the present-day global average of 14 °C (57 °F). The Mesozoic Era was dominated by Hot Houses, but there is evidence for a cooler period near the end of the Jurassic and the beginning of the Cretaceous when there was probably ice in the polar regions. Also there is evidence for a sharp decline in sea level between 126 and 128 million years ago, which suggests that there was glaciation in continental interiors at high latitudes.

The mid to late Cretaceous was probably one of the hottest periods in Earth's history (Brady et al., 1998). During the warm climate extreme 80 to 90 million years ago the average global temperature was probably between about 22 and 29 °C (71 and 84 °F), compared to the present 14 °C (57 °F). The deep ocean is estimated to have been about 15 °C warmer than today, and the atmospheric CO_2 content was much greater than today. Under these conditions there would be no ice or snow on Earth, and the sea level would be on average 250 meters higher than at present. At that time the Antarctic continent was almost at its present position but was covered with pine forests rather than the present ice sheet. In the Arctic Ocean the surface water temperature was 15 °C or more, which is the same as today off Maryland and France at latitudes between 35 and 45 °N (Jenkyns et al., 2004). Recent research indicates that the CO_2 content at that time was between 1,300 and 2,300 ppm compared with the 383 ppm today. Amazingly, in the tropical Atlantic the ocean surface temperature was about 42 °C (almost like a hot tub), or 14 °C higher than today (Bice et al., 2006).

This has profound implications for some of the scenarios concerning present-day global warming. Humans have never experienced this type of climate, and it is problematic whether we could survive under such conditions. All of human evolution has taken place in an Ice House. Some climate models of present-day global warming predict an extreme of 11 °C increase in average global temperature within 100 years if greenhouse gas emissions continue to rise. This would place us in the equivalent of the late Cretaceous climate extreme. It also implies that the current computer models that have used similar amounts of CO_2 in their simulations may have greatly underestimated the amount of warming. They produce substantially lower temperatures than those measured from the sediment cores.

Mass Extinctions

During the Paleozoic and Mesozoic Eras there have been five mass extinctions of plant and animal life (Figure 5.3).[1] They occurred at the end of the Ordovician 443 million years ago, the end of the Devonian 354 million years ago, the end of the Permian 248 million years ago, the end of the Triassic 206 million years ago, and the end of the Cretaceous 65 million years ago. At least two of these extinctions occurred very rapidly. The greatest mass extinction occurred at the end of the Permian when 84% of existing genera and 95% of existing species were exterminated. Most of these were marine organisms, but numerous species of land animals and plants also became extinct. There have been other large extinctions, but nothing comparable to the "Big Five."

The consequences of these events were absolutely devastating to life. Over the last 543 million years during the Phanerozoic Eon, 99.9% of all species that ever lived have been extinguished. However, each of the mass extinctions triggered a surge in evolution that resulted in large increases in biological diversity. Today there are at least a million species, 75% of which are insects.

---

[1] A mass extinction is defined as the extinction of at least 50% of existing species or 40% of existing genera.

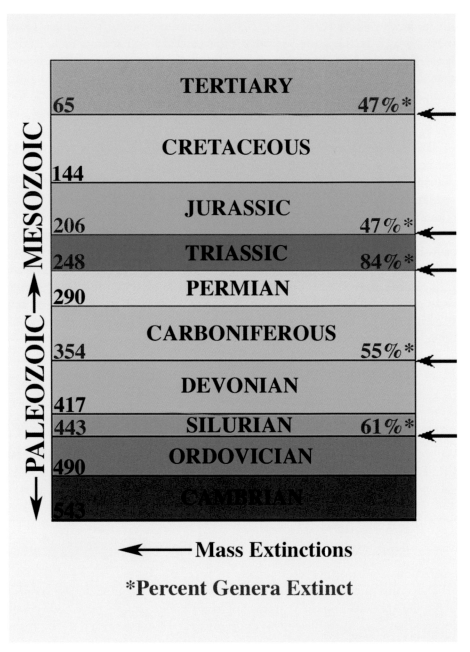

FIGURE 5.3 *This diagram shows the time of the five largest mass extinctions with respect to the geologic time scale. The red number on the right is the percentage of the genera that became extinct and the red number on the left is the time at which the extinction occurred in millions of years.*

We are now sure that the last mass extinction at the Cretaceous/ Tertiary (abbreviated to K/T) boundary was the result of extreme and almost instantaneous climate change caused by the large impact of an asteroid or comet. The estimated impact frequency for objects comparable to the K/T impact is about 1 every 100 million years, and there have been five mass extinction events in the last 540 million years. There is some geologic evidence that other mass extinction events may have been the result of impacts.

The recent discovery of shocked quartz and meteorite fragments in a breccia (a rock containing large angular fragments) at the Permian/Triassic boundary in Antarctica and China has been cited as evidence that this, the largest mass extinction, may have been the result of a giant impact. However, a recent study suggests that the extinction may have been due to a sudden climate change resulting from the injection of massive amounts of CO_2 into the atmosphere by a period of catastrophic volcanism that took place at that time. Computer models suggest that the dramatic rise in CO_2 caused temperatures to soar 10 to 30 °C higher than today. The warming would have had profound effects on the oceans by cutting off oxygen to lower depths and killing most of the marine life (Huey and Ward, 2005). The recent satellite discovery of an impact crater in Antarctica about four times larger than the K/T impact crater appears to be the same age as the Permian/Triassic boundary. This crater would be about 800 km in diameter and may have caused the release of massive amounts of CO_2 instead of volcanism. Additional studies are required to distinguish between these causes.

There are two compelling lines of evidence that the K/T extinction was the result of an impact. First, a large buried impact crater about 200 km in diameter was discovered as a result of oil exploration on the Yucatan Peninsula of Mexico. It was probably formed by the impact of an asteroid about 10 km in diameter. This crater has been dated at 65 million years, exactly the time of the K/T extinction. Second, wherever the K/T boundary is preserved there is a layer of impact ejecta that decreases in thickness with increasing distance from the crater. The layer was first discovered at the well-preserved K/T boundary near Gubbio, Italy, where it settled out of the atmosphere 65 million years ago. The layer is unusually rich in iridium, which is abundant in asteroids but rare in terrestrial rocks. It also has shocked quartz and high-pressure minerals, which are very strong indications of impact.

Computer simulations indicate that the impact itself did not cause the extinction, but rather the catastrophic climate changes that it produced. First, the impact must have ejected large amounts of material above the atmosphere to re-enter at relatively high velocities. This material would have heated the atmosphere to high temperatures and started forest fires that consumed large parts of the Earth. Second, the impact was in sulfur-rich sediments that would have provided a particularly strong reflective layer of dust in the atmosphere. This dust would have drastically decreased the solar insolation to the point where severe global cooling must have occurred. As a consequence, extensive glaciation would have occurred for about 10 years following the impact. Furthermore, the reduced sunlight would have severely inhibited photosynthesis, greatly disrupting food production. Eventually the sulfate aerosols would precipitate as acid rain, further disrupting food production. Also, large volumes of limestone were vaporized producing abundant atmospheric CO_2 that would result in global warming after the cold period. These conditions would certainly be "hell on Earth." They would have killed many species, first by raising the temperatures of the atmosphere to high levels for a short time, then plunging the Earth into a relatively short-lived ice age for a decade or so, followed by global warming from the injection of massive amounts of CO_2 into the atmosphere. These rapid climate extremes would surely result in an enormous mass extinction.

The K/T impact resulted in the demise of 47% of existing genera and about 75% of existing species, including the dinosaurs. A vacant ecological niche was created by this mass extinction event. It was filled by a great diversity of surviving species, particularly the mammals. Mammals were present during the Mesozoic Period when they were small shrew-like animals that probably burrowed under ground. Recently a Mesozoic mammal was discovered that was an amphibian beaver-like creature. The earliest mammal so far discovered is a rat-like creature about 210 million years old. However, mammals did not diversify very much during the remaining 145 million years of the Mesozoic Era. In fact, dinosaurs were dining on our earliest ancestors, and probably kept the mammal population suppressed.[2] With the end of the dinosaurs, mammals diversified rapidly to fill the vacant ecological niche until today they are the dominate species in terms of mass. We, in fact, owe our very existence to the impact that resulted in the K/T extinctions. If that impact had not occurred, dinosaurs would still be around today, eating mammals and relegating them to a rather minor part of the animal kingdom. It is doubtful that humans would have evolved in those circumstances.

It is possible that we are currently at the beginning of a sixth mass

---

[2] In China, a mammal fossil was found within the stomach of a dinosaur fossil. It apparently ate the mammal just before it died.

extinction—this time caused by humans. Ecologists estimate that, at the current rate of human-caused biological extinctions, as much as half of all species will be extinct in 100 years. Global warming is also beginning to cause extinctions.

Tertiary Climates

After the perturbations of the K/T impact, the climate in the Paleocene Epoch returned to hot conditions similar to those in the late Cretaceous (Figure 5.4). There was a brief exceptionally warm period at the end of the Paleocene about 55 million years ago called the Paleocene Eocene Thermal Maximum (PETM), shown in Figure 5.4. This event is important because it may have ominous implications for our current warming trend. Over a period of about 10,000 years at the end of the Paleocene Epoch, the Earth's climate and oceans warmed (Zachos et al., 2003). The atmospheric surface temperatures rose between 5 and 10 °C, and the deep-ocean and high-latitude surface waters soared by 8 to 10 °C. Bottom water temperatures rose by about 4 to 5 °C. In the tropical Pacific Ocean the sea surface temperatures rose about 4 to 5 °C (Tripati and

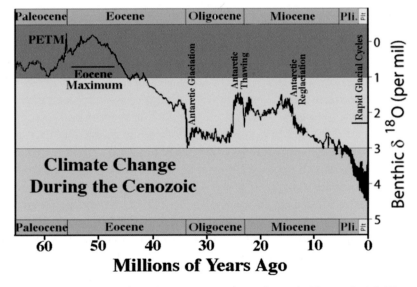

FIGURE 5.4 This diagram shows the variations in climate during the Tertiary Period. The curve is an indication of temperature variations from oxygen isotope variations in ocean sediments (the lower the value the higher the temperature). The spike labeled PETM is the Paleocene Eocene Thermal Maximum discussed in the text. The blue region is the present Ice House, while the red region is a Hot House. The yellow area is transitional between the two. The black horizontal bar marked "Rapid Glacial Cycles" is enlarged in Figure 5.5. (Modified after Robert A. Rohde at Global Warming Art at: http://www.globalwarmingart.com/wiki/Image:Instrumental_Temperature_Record.png)

Elderfield, 2005). Temperature proxies (climate indicators) from sediment cores near the North Pole indicate a water temperature of about 24 °C or the same as present–day subtropical ocean temperatures (Sluijs et al., 2006). Today the water temperature in the Arctic Ocean is about −2 °C. Fifty-five million years ago the CO_2 level was probably about 1,000–1,500 ppm. Disturbingly, climate models for 55 million years ago do not come close to simulating such warm waters even with CO_2 levels as high as 2,000 ppm. During that time many deep-sea species of foraminifera (microscopic ocean animals) became extinct, and numerous mammalian orders appeared, including primates.

This extraordinary event coincided with a drastic decrease in the $^{13}C/^{12}C$ isotope ratio of carbon reserves (Norris and Röhl, 1999), and is probably explained by the addition of isotopically light CO_2 to the ocean. The drastic increase in temperature at that time was probably due to the release of massive amounts of CO_2 or methane. Alarmingly, today human activity is releasing greenhouse gases 30 times faster than they were being emitted then (Zachos et al., 2003). Three mechanisms have been proposed for this release. One is a sudden change in ocean circulation causing a catastrophic release of about 1,200 to 2,000 billion metric tons of carbon in the form of methane trapped in marine sediments. About two-thirds of this release was completed in a few thousand years at most. A second possibility is the release of very large quantities of methane from continental areas. Fifty-five million years ago the world was very dry and large sustained peat fires may have released enormous quantities of methane to the atmosphere causing the thermal maximum (Katz et al., 1999). A third possibility is that massive volcanism from newly discovered 55-million-year-old vents under the Atlantic Ocean disturbed huge amounts of buried methane and CO_2 gas in marine sediments. The estimate is about 1,500 billion metric tons of carbon equivalent released in a geologically very short period of time. This climate extreme lasted at least 100,000 years. Today there is about 14,000 billion tons of methane locked up in marine sediments that could be released by some triggering mechanism (probably ocean temperature increases). There are also huge quantities of methane locked up in northern peat lands and this gas is now beginning to be released in some localities.

When the continents migrated to a distribution similar to today, the climate began to change (Figures 5.1 and 5.2). After an early warm period in the early Eocene, the Earth began to cool off about 45 million years ago, but not uniformly (Figure 5.4). In both the Arctic and Antarctic at that time icebergs were present and CO_2 levels were declining (Moran et al., 2006). They continued to drop until about 20 million years ago. There were two rapid cooling events, one at the Eocene/Oligocene boundary about 32 million years ago and another at 15 million years ago. A sea level change occurred near the end of the Eocene and beginning of the Oligocene. This suggests that glaciers were forming in Antarctica at that time and recent studies indicate that ice was forming in the Arctic in the mid-Eocene about 45 million years ago (Moran et al., 2006). Unlike today there were forested areas in Antarctica as late as 25 million years ago.

There was further cooling about 12 to 14 million years ago that initiated mountain glaciers in the Northern Hemisphere, and established a permanent ice sheet over east Antarctica. Finally, a sharp cooling between 5 and 6 million years ago produced a permanent ice sheet over all of Antarctica and established the current Ice House. Beneath a 0.5-meter debris cover in the ice-free Beacon Valley, part of the Dry Valley region of Antarctica, there is still relic ice greater than 8.1 million years old from the Miocene glacial epoch (Marchant et al., 2002).

The causes of the Tertiary general cooling are probably linked to continental drift that increased the amount of land at high northern latitudes. This made it easier for ice sheets to form, and resulted in various openings and closings of seaways changing the thermohaline circulation of the oceans. There were also decreases in the CO_2 levels.

Pleistocene Climates

The Quaternary Period is divided into the Pleistocene Epoch from 1.6 million years ago to 10,000 years ago, and the Holocene that is the current interglacial period from 10,000 years ago to the present.

The Pleistocene had repeated glacial and interglacial periods (Figure 5.5). During glacial periods, up to 32% of the Earth's surface was covered

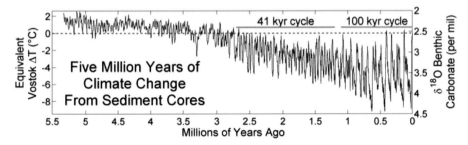

FIGURE 5.5 The onset and fluctuation of the Pliocene and Pleistocene ice ages are shown in this graph. The 41,000-year cycle began about 2.7 million years ago and was probably caused primarily by the Milankovitch cycles. The 100,000-year cycle began about a million years ago. Homonids started in the relatively cold period about 2.3 million years ago, and early Homo sapiens began during an interglacial period 400,000 years ago. The area below the dashed line is the current Ice House. See text for discussion. (Source: Robert A. Rohde at Global Warming Art at: http://www.globalwarming art.com/wiki/Image:Instrumental_Temperature_Record.png

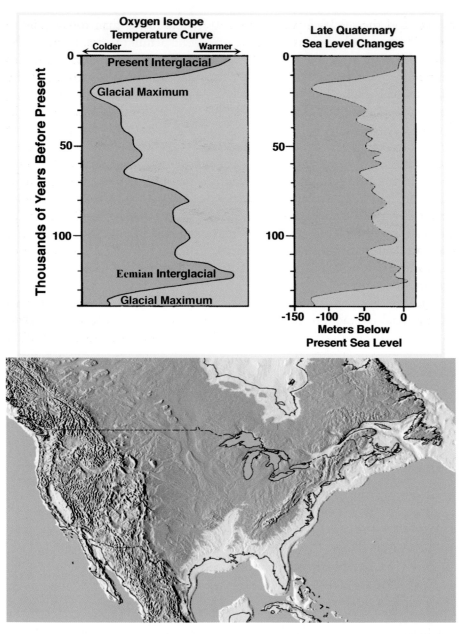

FIGURE 5.6 *This shows the change in sea level during the last glacial and interglacial periods. (Top) A diagrammatic temperature curve (left) and the change in sea level resulting from the deposition or melting of the ice sheets. (Modified from Van Andel, 1994, Fig. 4.12) (Bottom) This graphic shows the changes in sea levels for the United States region during the last ice age (line between light blue and darker blue), the present interglacial period (present shore line in black), and if all ice on Earth were to melt (line between light blue and brown). (Courtesy Tasa Graphic Arts, Inc.)*

by ice, and the sea level was about 120 meters lower than today (Figure 5.6). The great glaciations really began about 3 million years ago in the Pliocene when permanent ice sheets formed at high northern latitudes (Figure 5.5). These glaciations were probably initiated by closing of the Isthmus of Panama, changing the thermohaline circulation at that time. Successive glaciations became progressively more intense, culminating in the great glaciations in the latter part of the Pleistocene. Before about 850,000 year ago glaciations were separated by about 41,000 years, consistent with the period of axial tilt variations. During the past 850,000 years they have occurred about every 100,000 years, which resulted in major changes in ice volume of 50 to 60 million km^3, and accompanying sea level changes with amplitudes of about 120 to 140 meters (Figure 5.6). The cause of this change in glacial cycles from 41,000 to 100,000 years has been attributed to changes in the amount of atmospheric CO_2 (Berger et al., 1993), or to a major increase in the temperature contrast across the equatorial Pacific (Garidel-Thoron et al., 2005).

The insolation in the atmosphere varies in a cyclical manner due to gravitational interactions between the Earth, Moon, Sun, Jupiter, Saturn, and Venus that changes the motions of Earth. This periodicity is known as *Milankovitch forcing* after the person who discovered it. Although these perturbations hardly affect the average solar energy striking the Earth, they alter the seasonal and geographic distribution of incoming solar energy (insolation) by as much as 20%. The forcing is roughly the same as the 40,000-year glacial cycle but not the longer one. During the 100,000-year periodicity the size of the insolation changes are small. The longer cycle is more consistent with variations in the ellipticity of Earth's orbit. There is, however, a strong 413-year astronomical forcing that is not seen in the geological record. The orbital and obliquity variations of Earth are, at least in part, the cause of the ice ages, but there are probably other effects. These include major changes in the thermohaline circulation of the oceans, and cooler summers that resulted in the southward migration of vegetation to create favorable conditions for continental ice-sheet growth.

The Earth has experienced nine ice ages in the past 780,000 years. The last glacial period, called the Wisconsin in North America, covered large parts of North America including the northern United States and much of northern Europe and Asia (Figures 5.7 and 5.8). It lasted about 100,000 years from 117,000 to 10,000 years ago. It was not one continuous very cold period. Instead it consisted of a series of cool to very cold periods with the coldest occurring near the end of the glacial period about 21,000 years ago (Figure 5.8). At that time the global mean surface temperatures were probably only about 5.4 °C cooler than the mid-20th century

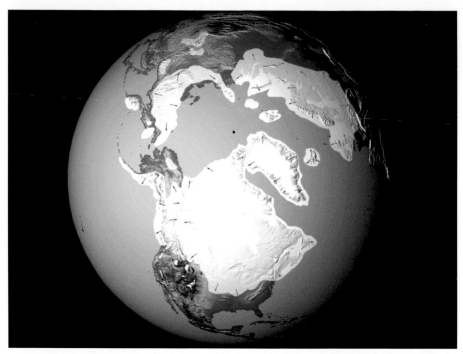

FIGURE 5.7 *This graphic shows the extent of the ice sheet that covered part of the Northern Hemisphere during the last ice age. The ice also covered the ocean next to the land ice. The black dot is the North Pole. (Courtesy Tasa Graphic Arts, Inc.)*

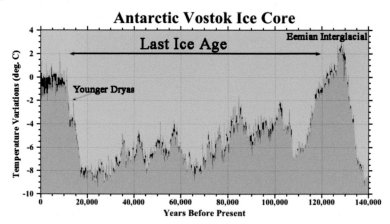

FIGURE 5.8 *In this diagram the temperature variations from the current global average (0) during the past 140,000 years is shown for the present interglacial (Holocene), the last ice age, and the last interglacial period (Eemian). The Eemian was mostly warmer than the Holocene because of differences in the Earth's orbital parameters. The temperatures are derived from proxies contained in the Antarctic Vostok ice core. (Vostok data courtesy Petit, J.R. et al., 2001, Vostok Ice Core Data for 420,000 Years, IGBP PAGES/World Data Center for Paleoclimatology Data Contribution Series #2001-076. NOAA/NGDC Paleoclimatology Program, Boulder CO, USA)*

(Ballantyne et al., 2005). However, there was larger average cooling at higher latitudes (~ 6 to $8\,°C$), and little or no cooling over large parts of the tropical oceans (Ruddiman, 2001). During some of the extreme glacial stages the air temperature plunged to $17\,°C$ below today's level (Bintanja et al., 2005). This glacial period was punctuated by very abrupt warming and cooling episodes. The period terminated rather abruptly about 12,000 years ago.

Antarctic ice core studies have shown that during glacial terminations there is a lag of about 800 years between the rise in temperature and the increase in CO_2. It takes about 5000 years for the warming to be complete, but only the first 800 years for the lag. Therefore, the first 500 years of warming were not caused by a CO_2 increase. However, the warming in the following 4200 years could be caused by CO_2. Some process, possibly related the Milankovitch cycles, causes Antarctica and its surrounding ocean to warm. This, in turn, causes CO_2 to be released about 800 years later (about the time of ocean overturning). At this time CO_2 further warms the planet causing even more CO_2 release. Consequently, CO_2 during ice ages should be regarded as a positive feedback to the warming. Model estimates indicate that CO_2 and other greenhouse gases cause about half of the full glacial to interglacial warming.

Although the length of ice ages averages about 100,000 years for the past 500,000 years, interglacial periods were much shorter; on average about 12,000 to 15,000 years long. Also, the 100,000-year cycles have resulted in the four mildest interglacial intervals in the past million years, including our own. The last interglacial period, called the Eemian, lasted 13,000 years between 117,000 and 130,000 years ago. The Eemian interglacial was generally warmer and drier in Europe and North America than our current Holocene interglacial period that may have been in part due to a higher orbital eccentricity at that time (Kaspar et al., 2005). This would have resulted in an excess of Northern Hemisphere solar insolation compared to today. Maximum temperatures reached $5\,°C$ warmer than today (North Greenland Ice Core Project Members, 2004). In fact, the extent of the Greenland ice sheet during much of this period was considerably smaller than today. This melting of the Greenland ice sheet alone may have raised the sea level several meters above its current level. In Antarctica it was about 3 to $5\,°C$ warmer than present. Unlike our present interglacial period, there was apparently one intense cool period in the middle of the Eemian interglacial about 122,000 years ago that lasted about 400 years. This event may have been the result of the incursion of fresh water through the Bering Strait during times of rising sea level and increased precipitation over the North Atlantic because of the Northern Hemisphere insolation maximum at that time. Of all the interglacial periods during the past 400,000 years, ours (the Holocene) appears to be one of the most climatically stable. It is now about 12,000 years long, which is about the average length of other interglacial periods. However, we may not be on the verge of another ice age. Recent ice cores from Antarctica have sampled an interglacial period (MIS 11) about 430,000 years ago that had warmer air temperatures, higher sea-surface temperatures, strong thermohaline circulation, high sea level, and greater coral reef expansion than the Holocene. At that time the Earth's orbital and obliquity (axial tilt) parameters were similar to our present interglacial period. That interglacial period lasted 28,000 years. If Earth's orbital parameters are crucial to the onset of ice ages then we may have at least 16,000 years or more before the next ice age. The one big difference is that the current atmospheric CO_2 content is higher than at any time in the past 650,000 years, and probably much longer.

ABRUPT CLIMATE CHANGES

One of the most startling scientific discoveries in recent years is that climates can change radically in a very short time period (National Research Council, 2002; Alley et al., 2003; Alley, 2004). In the early 20th century, former climate changes were believed to take tens of thousands of years to develop. In the 1950s, however, some scientists suspected that past climates had changed in only a few thousand years. The 1960s and 1970s saw scientists beginning to realize from different research that past climates could change radically within a few hundred years. As instrumentation and measurements became more sophisticated in the 1980s and 1990s it was discovered that radical departures in climate have taken place within a hundred years. In fact, it was found that the average temperatures over Greenland had sometimes risen an incredible 7 °C in less than 50 years. Today, with more extensive research and more accurate measurements, there is compelling evidence that severe climate change can occur in less than 10 years. For instance, drastic shifts in the entire North Atlantic climate have been observed within just five glacial snow layers, or only five years.

What is an Abrupt Climate Change?

An abrupt climate change occurs when the climate system is forced to cross some threshold that triggers a rapid transition from one climate regime to another. The forcing need not be radical. Feedbacks between different environmental components often result in non-linear responses that are very large compared to the original forcing. In fact, the forcing may be chaotic and so small that it is undetectable. A slow persistent forcing like current global warming could trigger an abrupt climate change with cataclysmic consequences for humans and other forms of life.

Evidence for Abrupt Changes

The best-documented abrupt climate changes occurred during the last ice age. They are spikes in the temperature and ice accumulation known as *Dansgaard–Oeschger events* after the Danish and Swiss glaciologists who discovered them in Greenland ice cores (Figure 5.9). The transitions into the cold phases of these events coincide with what are known as *Heinrich layers*, named after the scientist who first discovered them. These are layers in Atlantic Ocean sediment cores that contain unusually high concentrations of ice-rafted debris (Hemming, 2003). This glacial debris is sediment that was contained in floating icebergs or sea ice, and then settled to the seafloor when the ice melted. The layers appear to have a near-global signature that coincides with other events such as changes in sea-surface

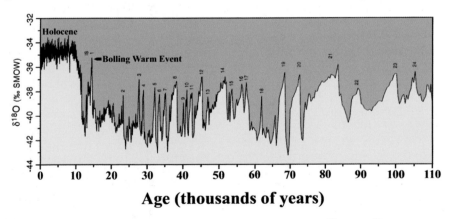

Age (thousands of years)

FIGURE 5.9 *This shows the variation in air temperatures from the ^{18}O and ^{16}O ratios in a Greenland ice core for the past 110,000 years. The numbers indicate what are called Dansgaard–Oeschger warm events that interrupt the otherwise cold conditions of the Ice Age. These spikes are abrupt climate changes that occurred during the last ice age and indicate how unstable the climate was during that period. The Bolling Warm Event resulted in a sea level rise of about 20 meters in a period of about 400 years. (Modified from Kennett et al., 1997)*

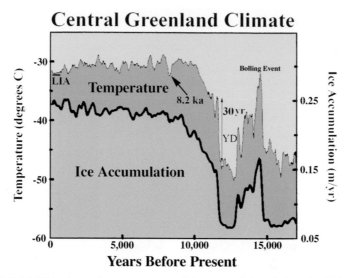

FIGURE 5.10 *This shows the temperature and ice accumulation during the Holocene and the termination of the last ice age. The end of the Younger Dryas (YD) occurred 11,600 years ago and only took about 30 years. LIA is the Little Ice Age, and the 8.2-ka event is the cold period that occurred 8,200 years ago. (Modified from Alley et al., 1993)*

temperature and sea level. The origin of the Heinrich events in the North Atlantic was probably massive discharges of icebergs from the *Laurentide ice sheet* through the Hudson strait. These events were followed by a slowing of the rate of the North Atlantic thermohaline circulation producing a colder climate. When the rate subsequently accelerated, the climate warmed abruptly.

A particularly noteworthy abrupt climate event is a striking shift into and out of a very frigid period about 12,000 years ago. It is called the Younger Dryas event after the Arctic flower *Dryas octopetala* whose pollen indicates cold tundra conditions. It lasted about 1,300 years (Figure 5.10). Figure 5.11 shows 100-year enlargements of the ice accumulation rate in Greenland at the beginning and end of three events.[3] The end of the Younger Dryas took only 40 years, but most of it occurred within 10 years. This event began with a rapid warming that melted part of the Laurentide ice sheet and produced a huge lake in Canada called Lake Agassiz (Perkins, 2002) which was about 134,000 km^2 in area and about 200 meters deep. A water volume of about 9,500 km^3 was released suddenly from this lake into the North Atlantic to shut down the thermohaline circulation and plunge the world back into the ice age. At

---

[3] Ice accumulation rates are proxies for warmer or colder climates.

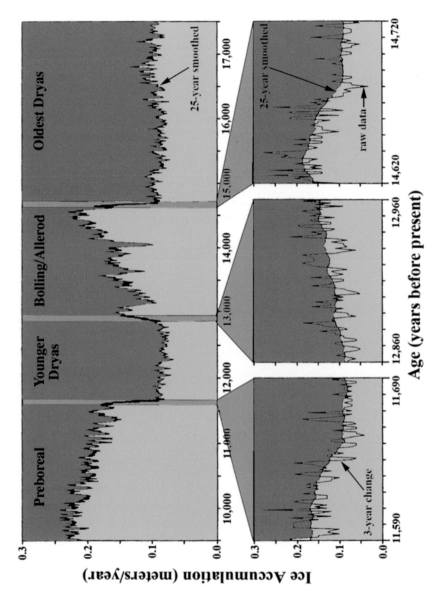

FIGURE 5.11 This diagram shows enlargements of several abrupt climate changes. The interval between tick marks in the enlargements is 10 years. (Modified from Alley et al., 1993)

that time humans were occupying the Middle East along the Fertile Crescent. They were hunter–gathers collecting wild nuts and grains, but the return of cold conditions brought severe drought to the region and, consequently, there was not enough wild grain to feed them. They, therefore, started planting their own grain and nut trees, which was the beginning of agriculture by humans. Thus, civilization was born from a drastic climate change (Fagan, 2004). As the climate warmed at the end of the Younger Dryas, agriculture flourished and permanent buildings became established that eventually led to the great city-states. Since then the climate has been fairly stable up to the present time.

All of the above abrupt changes have taken place during ice ages, probably because the ocean–atmosphere system is particularly unstable during glacial periods. However, there is at least one rapid warming event that happened during a Hot House about 55 million years ago. This event was described in "Tertiary Climates" (page 61). It was apparently the result of a rapid change in the thermohaline circulation that resulted in the catastrophic release of methane in seafloor sediments. Such events may also have happened in other Hot Houses.

A radical abrupt climate change requires four conditions: (1) a trigger that abruptly pushes the climate into a new regime; (2) feedback mechanisms that intensify the climate; (3) a process for spreading the change over the Earth, or at least over a hemisphere; and (4) a process that causes the climate change to persist for long periods (decades to centuries). We know many triggers, as discussed above, but only a few are capable of triggering a rapid change. Feedback mechanisms for amplifying the climate are known to some degree, but there may be others of which we have no knowledge. The mechanisms that spread a climate change across the globe are not well understood.

Possible Causes

Two main causes have been suggested for abrupt climate changes. One involves rapid changes in the thermohaline circulation to cool the climate, and the other involves the catastrophic release of methane to warm the climate. As explained in Chapter 4, changes in the thermohaline circulation involve the massive addition of fresh water to the North Atlantic from large glacial lakes due to melting glaciers. Melting glaciers and/or large increases in northern river discharge rates can introduce large amounts of fresh water into the North Atlantic that changes the salinity of the ocean so that the denser warm current plunges down at lower latitudes shutting down the transport of ocean heat to high latitudes. This could lower the surface temperatures by as much as $6\,°C$ in the Northern

Hemisphere, plunging it into an exceptionally cold climate or, under the right conditions, into an ice age. This appears to be unlikely today because of the exceptional warming due to the high atmospheric CO_2 content.

Another hypothesis for the abrupt warming events is the periodic, catastrophic release of methane, three sources of which have been proposed. One is the catastrophic emission of methane caused by instabilities of methane hydrate contained in marine sediments on continental slopes. These are areas where methane hydrate is stable under the proper temperature and pressure conditions, but warming of these areas and/or lowering of the pressure due to lower sea level would destabilize the methane hydrate and catastrophically release massive amounts of methane into the atmosphere. Methane, of course, is a powerful greenhouse gas (23 times more powerful than CO_2) that would rapidly raise the atmospheric temperature. Another source is methane release from submarine natural gas seeps that are perturbed by lower sea levels and reduced pressure on the seeps that allows more gas to be released to the atmosphere. Another possibility is the massive release of methane from spreading wetlands during warmer periods within a glacial epoch. Recent studies of the hydrogen isotope ratios of methane in ice cores suggest that it originated from wetlands rather than methane hydrates (Sowers, 2006).

Although abrupt climate changes are now well established within the climatology community, they are not widely appreciated by many scientists, and particularly by social scientists and policy makers. Fortunately, severe and prolonged abrupt climate changes have never occurred since civilization began some 12,000 years ago. Abrupt climate changes may occur for many reasons, but it is certainly conceivable that human forcing by the emission of greenhouse gases could push us over a threshold into an abrupt and radical climate change. Today the greenhouse gas content of the atmosphere is greater than at any time in at least the past 650,000 years. We may be living on borrowed time. Can you imagine the global political, social, and economic disruption caused by a drastic climate change that took place in only 5 or 10 years? This is truly our worst nightmare.

SUMMARY

The geologic past has had numerous natural climate changes due to a variety of causes. In general they can be grouped into what are called "Hot

Houses" and "Ice Houses." In a Hot House there was little or no ice on Earth, and the average global temperature was considerably higher than today, in some cases over 24 °C. Hot Houses appear to have been more common than Ice Houses. Ice Houses are cold periods, usually alternating glacial and interglacial periods. Today we are in an interglacial period of an Ice House, but our current warming may put us in a Hot House in the next 100 years.

One of the most startling recent discoveries is that climate change can happen in short periods of time, in some cases 10 years or less. These abrupt changes occur when the climate system crosses some threshold, possibly the rapid release of greenhouse gases or a rapid change in the thermohaline circulation. This has resulted in global average temperature changes of as much as 8 °C in less than 20 years. As the causes of these abrupt changes are not well understood, we must be aware that the current rapid greenhouse gas increase and rapid warming could trigger an abrupt climate change with devastating consequences.

6

THE HEAT IS ON

T HERE is now good evidence that our present interglacial period has experienced some climate shifts that have affected the course of history in different ways. New studies indicate that these climate shifts were mostly regional rather than global and less severe than previously thought. Aside from a sudden cooling 8,200 years ago and short-lived cooling events caused by very large volcanic eruptions, the current global warming represents the most radical global climate shift in probably the last 1,800 years and definitely in the past 400 years.

THE HOLOCENE: OUR PRESENT INTERGLACIAL PERIOD

The Holocene extends from 10,000 years ago to the present, and encompasses almost all of civilization. It has had a remarkably stable climate compared to the last ice age and even other interglacial periods. There have been three relatively strong cold periods in the Holocene. The best known was a cold period 8,200 years ago called the 8.2-ka event (Figure 6.1). This period lasted about 300 years with surface temperatures in the Atlantic region some 1.5 to 3 °C cooler than present. The event was probably caused by a sudden drainage into the North Atlantic of the reformed Canadian glacial lake Agassiz that had merged with the new lake Ojibway. The combined area of these lakes was about 841,000 km². The release of enormous quantities of fresh water through Hudson Bay probably altered the North Atlantic thermohaline circulation to temporarily stop or reduce the flow of warm water to high latitudes, causing temperature drops in parts of the Northern Hemisphere. The outflow of water into the Atlantic Ocean caused the sea level to rise about 0.5 meter. Recent computer models indicate that the effects were much milder that previously thought (LeGrande et al., 2006). Temperatures in Greenland and the North Atlantic had the largest decreases with slightly less cooling in parts of North America and Europe. The rest of the Northern Hemisphere had almost no temperature change, and the Southern Hemisphere was largely unchanged, consistent with Antarctic ice core data (Figure 6.1). Initially the ocean circulation decreased by almost half but it rebounded in about 50 to 100 years. These simulations suggest that a current disruption of the thermohaline circulation would not be as drastic as once thought, particularly under the current much warmer conditions. There were two other cool events that show up in the Greenland ice cores at 1,200 and 4,750 years ago (Figure 6.1).

Determining past global average temperatures from climate proxies is very difficult the farther back in time one goes. A new study by the US National Academy of Sciences (National Research Council, Committee on Surface Temperature Reconstructions, 2006) has attempted to access the accuracy of determining global average temperatures during the past 2,000 years. They find with a very high level of confidence that the last few decades of the 20th century up to the present were warmer than any comparable period in the past 400 years. For surface temperatures from AD 900 to 1600 less confidence can be placed on the data, but many locations were warmer in the past 25 years than during any other 25–year period since 900 AD However, there is very little confidence about *average global temperatures* prior to 900 AD because of the scarcity of proxy data for that time frame. With that in mind we will examine the best estimates of at least local temperatures during the past 5,000 years. The best data are for the past 400 years.

"Small" Climate Events

There are three relatively "small" climate events that have been recognized primarily in Europe (small compared to the climate changes

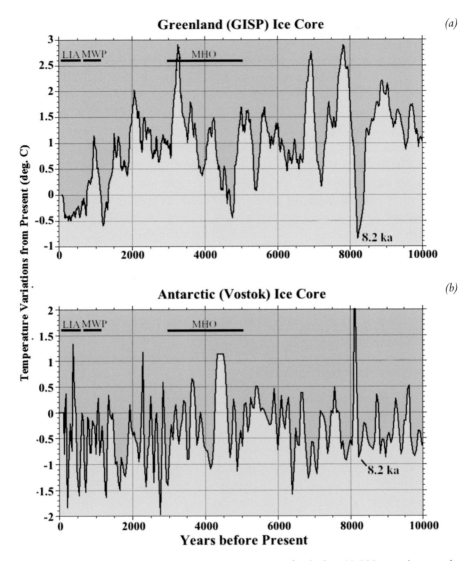

FIGURE 6.1 *Comparisons of the variations in temperature for the last 10,000 years between the Arctic and Antarctic from ice core data. The Little Ice Age (LIA), the Medieval Warm Period (MWP), the Middle Holocene Optimum (MHO) and the 8.2-ka event do not show up in the Antarctic record suggesting that they were confined to the Northern Hemisphere. Even the Middle Holocene Optimum in Greenland has two cool periods. (Vostok data courtesy Petit, J.R. et al., 2001, Vostok Ice Core Data for 420,000 Years, IGBP PAGES/World Data Center for Paleoclimatology Data Contribution Series #2001-076. NOAA/NGDC Paleoclimatology Program, Boulder CO, USA)*

discussed in the last chapter). These climate changes are usually grouped into periods that have been termed the "Mid-Holocene Optimum," the "Medieval Warm Period," and the "Little Ice Age." However, there is much confusion about the ages and duration of these periods as well as the size of the climate shifts and their locations. There is now good evidence that at least the Medieval Warm Period and the Little Ice Age were regional events mainly confined to Europe and the Atlantic Seaboard with some effects in other parts of the Northern Hemisphere but probably not the Southern Hemisphere. None of the three climate events, including the 8.2-ka event, appears in the Vostok ice core data (Figure 6.1(b)) as recorded in the Greenland ice core (Figure 6.1(a)). In fact, the stated duration of these events do not coincide very well with their apparent signatures in the Greenland ice core data (Figure 6.1(a)). In marked contrast, the current warming is of global extent. However, as small as these past climate changes were, they had significant impacts on civilization (Fagan, 2004; Linden, 2006). There were other abrupt short-lived (a few years) decreases in global temperatures due to very large volcanic eruptions and, at least in some cases, they also changed history.

The Mid-Holocene Optimum

The Mid-Holocene Optimum is generally considered to have lasted about 2,000 years from 1000 to 3000 BC, although there is still some uncertainty in these dates. During this period the average summer temperatures in the northern mid-latitudes of Europe and North America are thought to have been about 2 to 3 °C warmer than present summer temperatures. In Europe trees grew farther north and extended to higher altitudes than at present. At lower latitudes the Monsoon was much stronger with more rainfall in south Asia, the Near East and Africa. These conditions probably facilitated the spread of agriculture from its beginnings in the Near East, although there is considerable uncertainty regarding the relative global, annual mean warmth at this time. The cause of this warming is also not well understood. Unlike today, the atmospheric CO_2 content was about the same as other interglacial periods. It may have been a time when the westerly wind circulation was particularly strong and brought warm, wet conditions to high northern latitudes. Also the orbitally induced insolation changes may have favored warmer high-latitude summers but cooler winters and slightly cooler tropics. In fact, recent climate models suggest that annual and mean surface temperatures actually were cooler than those in the mid-20th century, even though northern mid-latitude summers were probably warmer (Kitoh and Murakami, 2002). This period has sometimes been used as an example of possible climatic conditions if present global warming continues. However, today's warming is global rather than regional, and the current temperature rise is greater than at any time in the past 10,000 years. Furthermore, we are not sure that global temperatures at that time were much higher than those at present.

General Cooling

Over the next 4,000 years northern mid-latitude summer temperatures appear to have cooled. This period is sometimes referred to as the "Neoglacial" because it contains periods of glacial advance (and retreat) of northern mid-latitude mountain glaciers. This period was very far removed from a full glacial period in the Pleistocene, and had at least one warm period. There was, however, a markedly cooler period from about 535 to 545 AD that appears to have changed at least European history. It marked the beginning of the Dark Ages, including the period of the Black Death—an event that may have resulted from an early gigantic eruption of Krakatoa volcano. Ice core data show a spike in the sulfuric acid deposition in 535 AD indicating such an event. In this case, a relatively minor cooling event (far removed from an ice age) appears to have changed European history dramatically (Fagan, 2004).

The Medieval Warm Period

Another warm period, which lasted about 400 years from 900 to 1300 AD, occurred during the European Medieval era and is appropriately called the Medieval Warm Period. There is much uncertainty about how warm and how widespread this warming was. None of the reconstructions indicates that temperatures were warmer during medieval times than during the previous few decades. It certainly occurred in Europe and the Americas, and it is recorded in the Greenland ice cores (Figure 6.1(a)), but it is manifestly different in other parts of the world. For example, there were notably wetter springs on the Tibetan Plateau from 929 to 1031 but apparently no temperature anomalies. The summer temperatures in Europe appear to be only about 0.5 °C above the long-term average between 1200 and 1350 AD, and winter temperatures were less extreme. Crops were grown farther north in Scandinavia and England than today, but there is no evidence for warmer temperatures in many other areas, such as the Urals. During this period the Norsemen established settlements in Iceland, Greenland, and North America. At that time temperatures in Greenland are estimated to have been 2 to 4 °C higher than today. In the Americas there were severe long-term droughts, hunger, and illness that led to the collapse of two civilizations; the Anasazi in the American west, and the Mayan culture in Central America. However, there is no evidence for this climate anomaly in Antarctic ice core data (Figure 6.1(b)), although there may have been some effect in the Southern Ocean (Goosse et al., 2004).

The Little Ice Age

The last climate change before the present warming is inappropriately called the Little Ice Age. The Pleistocene ice ages were many times worse. It may have lasted about 500 years between 1350 and 1850 AD (Figure 6.1(a)), but there is considerable uncertainty concerning its duration. Recent glacial studies indicate that it reached its maximum intensity about 1800 (Oerlemans, 2005). This was not just one 500-year period of cold temperatures; it was a time of fluctuating periods of hot summers and bitter cold winters (Figure 6.2). Between 1634 and the end of the 17th century the winter temperatures were 0.5 °C colder than subsequent centuries. On average, temperatures were about 1.3 °C colder than those of the Medieval Warm Period. In fact, the coldest century in the last 2,000 years was the 17th, with an average temperature that was only 0.5 to 0.8 °C below the 1961–1990 average. Although this temperature difference is relatively small, it had profound effects on European history. The winters were exceptionally cold in the 1690s when there were widespread crop failures. Glaciers advanced, beginning in the mid-13th century and destroyed small villages at high altitudes in the Alps and glacial advance continued into the 16th century. In 1589 a glacier blocked the Saas valley and its river in Switzerland, resulting in the formation of a lake that eventually broke through the ice and caused extensive flooding. In Chamonix, France, land and property were destroyed by glacial action. In North America glacial advances occurred from 1711 to 1724 and from 1835 to 1849. The more severe winters had profound impacts on the agricultural, economic, and political structure of Europe where famine and pestilence were common. During this period the Norse people abandoned their settlements in Greenland and North America. The Little Ice Age had changed the course of European history. One anthropologist (Fagan, 2004) called it "The mother of all history-changing events."

The cause of the Little Ice Age is not well understood. One possible contributing factor may have been the Sun's irradiance (Lean et al., 1995). The depths of the Little Ice Age coincide with the Maunder Minimum from 1645 to 1715 when sunspots were extremely rare (small numbers of sunspots correlate with low solar irradiance). There was another shorter

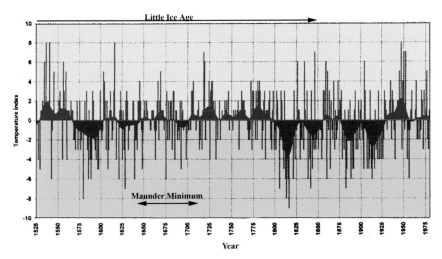

FIGURE 6.2 *This shows the temperature index (a measure of temperature variations) of Switzerland from 1525 to about1977. Also shown is most of the period of the Little Ice Age and the Maunder Minimum in solar irradiance. Although there were cold periods there were also warm periods, at least in Switzerland and the Maunder Minimum only occurred over a very small part of the Little Ice Age. (Modified from Burroughs, 1997, Fig. 2.9. Data from Pfister's paper in Bradley and Jones, 1995)*

time of low sunspot abundance from 1795 to 1825 known as the Dalton Minimum, and although these times of low solar irradiance may have been a contributor to the Little Ice Age, they are not enough to account for the entire length or magnitude of this cold period. However, there may have been positive feedbacks such as the increased area of ice and consequently greater reflection of radiation back to space. To make things even worse, there were three major volcanic eruptions in 1808 (not located), 1818 (Tambora, Indonesia), and 1883 (Krakatoa, Indonesia) that lowered the global temperature even more for short periods of time. The Krakatoa eruption lowered the global temperature by about 1 °C for several years, but both the 1808 and Tambora eruptions lowered the temperature even more, maybe about 1.5 °C. The Tambora eruption cooling in 1818 is known as "the year without a summer" when there were July frosts in New England.

Complex Patterns

The climate reconstructions for the past 2,000 years have led to a simplistic picture of a Medieval Warm Period and a Little Ice Age. Instead, the records of climate variability indicate much more complex patterns of past regional variations that rarely coincide with the actual patterns of hemispheric or global average variations (Mann and Jones, 2003; Jones and Mann, 2004). They are probably biased due to emphasis on one part of the world such as the North Atlantic/Europe region. Figure 6.2 shows inconsistencies in Europe where the temperature variations in Switzerland do not closely match the duration of the Little Ice Age, and show numerous warm summers. It is probably better to view the climate changes during the last 2,000 years in terms of cool and warm centuries in various parts of the world. For example, the early 19th century was cool in North America. In Europe the 16th, 17th, and 19th centuries were cool, but the 18th century was warm. Eastern Asia had a cool 19th century, and there was a cool period in the tropics from 1650 to 1750. During the Little Ice Age there was a discernible warm period and during the Medieval Warm Period there was a cool period. In other words, there was considerable climate variability throughout the past 2,000 years, but most of that variability appears to have occurred regionally in the Northern Hemisphere.

Although the climate changes that occurred during the past 1,800 years were small compared with earlier climate changes, they had profound effects on civilization at a time when the human population was small and the global environment was in much better condition than it is today. However, at that time civilization did not have the technology of today and it was almost always in turmoil, so small to moderate regional climate changes could disrupt societies. Large volcanic eruptions, in particular, lowered the global temperature greatly for short periods of time. Although short, these events greatly disrupted crop production and other societal characteristics, and caused, for example, a spread of disease that greatly affected societies in various parts of the world.

PRESENT GLOBAL WARMING

That global warming is happening now is not in doubt. Numerous studies have conclusively indicated that global warming is a fact and that it is a global process unique to at least the last 400 years and probably the past 1,000 years, if not longer. It strongly coincides with the Industrial Revolution and the rapid growth of the human population.

Unique Event

Smoothed (40-year-average) temperature reconstructions from climate proxies for the past 1,800 years are shown through 1995 in Figure 6.3

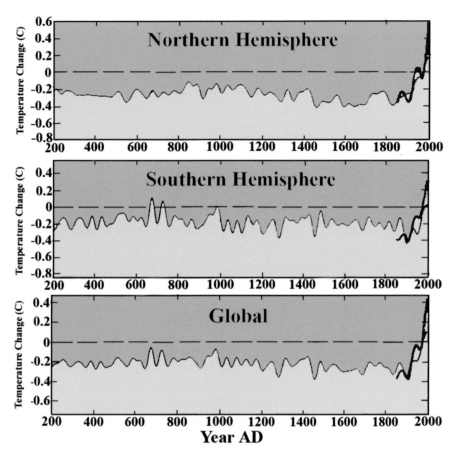

FIGURE 6.3 *This shows the temperature anomalies from climate proxies (thin wavy line) and observations since 1880 (red line) over the past 800 years. The dashed line is the 1961–1990 global average temperature. The current warming is unprecedented during that time. See text for explanation. (Modified from Jones and Mann, 2004, Fig. 5)*

(modified from Jones and Mann, 2004).[1] The red line is the instrumental record from about 1850 to 2003. The temperatures are relative to the

---

[1] McIntyre and McKitrick (2005) have challenged the original Mann et al. (1998, 1999) papers based on statistical tests and maintain that there is no difference in the past compared to today. However, Huybers (2005) has strongly criticized that analysis and showed that the Mann et al. results are basically correct. A re-analysis by Moberg et al. (2005) and Osborn and Briffa (2006) also basically support Mann et al. and the more recent Jones and Mann (2004) results. McKitrick is a senior fellow at the Fraser Institute, a free-market oriented Canadian think tank that received $60,000 from ExxonMobil in 2003. Both McIntyre and McKitrick are listed as "experts" for the George C. Marshall Institute, which has received $515,000 from ExxonMobil since 1998.

1961–1990 mean Northern Hemisphere temperature (dashed line). The reconstructions for the Southern Hemisphere, where the data are relatively sparse, have a greater uncertainty than the Northern Hemisphere. The scarcity of precisely dated proxy evidence for temperatures before 1600, especially in the Southern Hemisphere, is the main reason that there is less confidence in earlier global reconstructions. On the other hand, confidence in large-scale reconstructions is strengthened by the fact that the proxies generally exhibit strong correlations with local environmental conditions.

Over the past 2,000 years, prior to the marked warming in the 20th century, the temperature variability is relatively small. In North America on average the coldest and warmest years are less than 1.5 °C, and decadal differences are less than 1.0 °C. The maximum amplitude change has been about 1 °C. Although there were definitely warmer and cooler periods in the past 2,000 years, the 20th century has experienced the largest most rapid **global** temperature increase of any century in the last 2,000 years. In a recent study of the past 1,200 years in the Northern Hemisphere (Osborn and Briffa, 2006), the unique temperature extremes of the current epoch have been confirmed. The study finds warming from 890 to 1,170 conciding with the Medieval Warm Period and cooling from 1580 to 1850 that coincides with the Little Ice Age. However, the current global warming is far greater than any warming in the last 1,200 years. There is no question that during the past 15 years the temperature has risen more rapidly and to a higher level than in at least the past 2,000 years (Jones and Mann, 2004; Moberg et al., 2005). A study of world glaciers suggests that the present global warming began slowly about 1860 (Oerlemans, 2005). However, the Industrial Revolution started over 100 years earlier than that, and we were burning coal and cutting down forests long before the start of the Industrial Revolution. The one firm finding is that the current warming is greater than at any time in the past 400 years.

Recent studies of the temperature changes of the four main components of the Earth's climate system indicate that the entire Earth has warmed appreciably during the 20th century (Levitus et al., 2001; Beltrami et al., 2002; Huang, 2006). The Earth's continents, oceans, atmosphere, and cryosphere (Earth's ice) are all effected by the climate and, in turn, affect the climate. All of these systems have warmed during the 20th century, which demonstrates that the warming is indeed global in scope, and very different from previous Holocene climate changes (Table 6.1).

The oceans have warmed much more than any other Earth system. In fact, the world oceans are responsible for about 84% of the total increase of heat content of the Earth system between 1955 and 1998 (Levitus et al., 2005).

TABLE 6.1 *Earth's heat gain during the last 50 years*

Earth system	Energy heat gain (10^{21} Joules[a])
Oceans	145
Continents	10.4
Cryosphere (Ice)	8.1
Atmosphere	6.6

[a] A unit of heat energy.

The warming of the oceans cannot be explained by natural internal climate variations or by solar irradiation. Computer simulations using natural variations, solar, and volcanic aerosol forcings were unable to reproduce the observed warming and its vertical structure in the upper 700 meters of oceans around the world. Only simulations using human–caused emissions of greenhouse gases could replicate the observed ocean warming and its complex vertical structure (Barnett et al., 2005). Since the findings were the same for two different climate models, the results were not significantly affected by differences in the model formulations. Observed ocean warming is additional strong evidence that humans are causing global warming.

Figure 6.4 shows the amount of forcing for the main forcing factors that would affect the climate during that time period—for example, green-

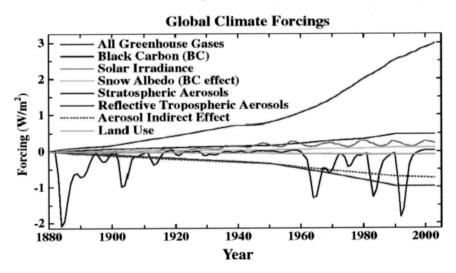

FIGURE 6.4 *This diagram shows the radiative climate forcing for various factors since 1880. The only significant long-term forcing is from greenhouse gases and aerosols. The stratospheric aerosols are from large volcanic eruptions like Krakatoa in 1883 and Pinatubo in 1991. (Courtesy James Hansen, NASA Goddard Institute for Space Studies, New York, NY).*

house gases, solar irradiation, volcanic eruptions, and aerosol emissions. It clearly shows that only the rapid rise in greenhouse gases correlates with the rapid rise in temperature. In fact, the rise in human-emitted aerosols should have cooled the Earth, but it was not sufficient to compensate for the warming caused by the rise in greenhouse gases. The amount of greenhouse gas forcing may have been greatly underestimated because of the relatively large aerosol content of today's atmosphere. There were apparently some temperature rises as high as $6\,°C$ during the ice ages with only a similar CO_2 increase observed today. These observations strongly indicate that human emissions of greenhouse gases are the primary cause of the rapid rise in temperature in the 20th century. There is also other compelling evidence, presented in the next chapter, that humans are the primary cause of global warming.

We are currently experiencing the largest and fastest global temperature rise that began slowly in 1860, but dramatically increased in the 20th century. It has greatly accelerated during the past 25 years. The temperature changes are usually compared to the average temperature between 1951 and 1980, although other years have been used. They are referred to as *temperature anomalies.*

Rising Temperatures

Figure 6.5(a) is the global annual mean and 5-year mean temperature anomalies from 1850 (when the worldwide instrument record began) to 2006. Both the annual and 5-year mean temperatures show the normal variations in temperatures, but the rapid rise in temperature during the past 25 years is unprecedented in the last 10,000 years. The graph also shows that 2005 was the hottest year on record. (The temperature extreme in 1989 was during an extreme El Niño year when global temperatures are unusually high.) Figure 6.5(b) is a similar plot for the Northern and Southern Hemispheres. The lower temperature in the Southern Hemisphere is because that hemisphere has a much larger area of oceans which tend to cool the region. The five hottest years on record all occurred between 1995 and the present. The two hottest were 1998 when the temperature reached $0.69\,°C$ above the 1951–1980 average, and $0.65\,°C$ in 2005 (Goddard Institute for Space Studies). The reason for the 1998 high temperature was the warming effect of a very strong El Niño event which boosted the temperature about $0.2\,°C$. Warm waters associated with El Niño increase the global temperature. When this is taken into account, 2005 was the warmest. The winter of 2006–2007 (December, January and February) was the warmest on record, indicating global warming is accelerating. The global temperature rise has now been $0.6\,°C$

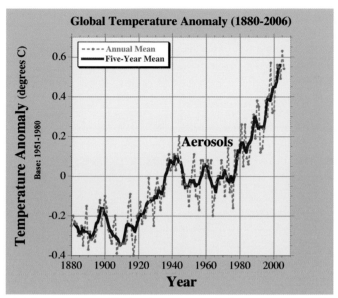

FIGURE 6.5a The annual mean (blue) and 5-year mean (red) temperature anomalies from 1880 to 2006 are plotted on this graph. The increase in temperature during the past 25 years has been remarkably rapid with 2005 the hottest year on record. The relatively flat region between 1942 and 1976 is due to the cooling effects of aerosols (See text for discussion). The drop in temperature for 1991–92 was due to the Mt. Pinatubo volcanic eruption. (Data courtesy NASA Goddard Institute for Space Studies, New York, NY.)

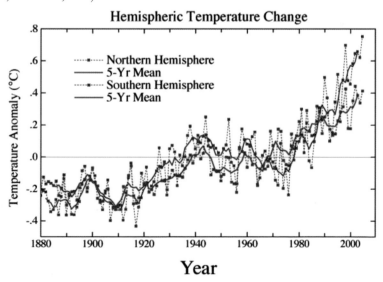

FIGURE 6.5(b) Plot of the hemispheric annual mean and 5-year mean temperature anomalies for the Northern and Southern hemispheres. The Northern Hemisphere has warmed more and faster than the Southern Hemisphere. See text for explanation. (Courtesy NASA Goddard Institute for Space Studies, New York, NY)

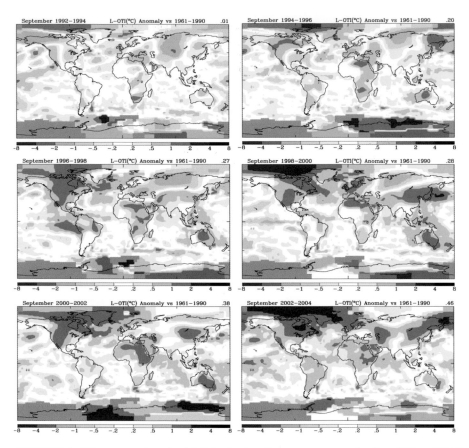

FIGURE 6.6 These maps show the global distribution of temperature anomalies from 1992 to 2004. The yellows and reds are positive (warm) anomalies while blues are negative (cool) anomalies. They show that the greatest amount of heating has occurred in the Arctic regions. (Courtesy of the Goddard Institute for Space Studies, New York, NY)

for the past 30 years and 0.8 °C in the last century. Up to 1975 in the 20th century there was slow global warming with large fluctuations, but subsequently there has been rapid warming of almost 0.2 °C per decade. In the past 155 years, 10 of the 11 warmest years have occurred between 1995 and 2005. Some people have argued that the observed rise in temperature is due to urban "heat island" effects near weather stations. This is certainly incorrect. The greatest warming, as predicted, is occurring in remote regions such as central Asia and Alaska. Also the largest areas of surface warming are over the ocean that is far removed from urban areas. In addition, continental surfaces have warmed about 0.5 to 1 °C in the past century. Figure 6.6 shows maps of the surface temperature anomalies in 2-year averages from 1992 to 2004.

There is a fair amount of natural variability but almost all of the warming occurred in the 20th century during two periods, 1910 to 1945 and 1976 to the present. The temperature rose by 0.37 °C between 1925 and 1944, and by 0.62 °C between 1978 and 2005. Presently, the average global temperature is about 14 °C in the Northern Hemisphere and 13.4 °C in the Southern Hemisphere (Jones et al., 1999).

The warming from about 1920 to 1945 and the fluctuations between 1946 and 1976 did not occur at all latitudes. For example, Arctic temperatures have been increasing since the 1960s. The reason for the fluctuating portion of the temperature curve between 1942 and 1976 is due to the increased emission of aerosols and particulates from coal-fired power plants and a major volcanic eruption in 1963 (Mt Agung, Indonesia) that cooled the Earth between 1964 and 1968. Aerosols and particulates cool the climate by reflecting sunlight back to space (Figure 6.4) and during this period the aerosols and particulates were offsetting the heating by greenhouse gases resulting in a more-or-less constant global average temperature. However, in the 1970s "scrubbers" were mandated for coal-fired power plants. After about 1975 the SO_2 emissions from the United States and Europe began a steep decline (see Figure 7.3) which lowered the aerosol and particulate emissions but still allowed the emission of CO_2, resulting in the beginning of the rapid temperature increase. Also at this time the CO_2 emissions were increasing at a very rapid rate, offsetting the cooling effects of aerosols (Figures 7.5 and 7.8). This is still happening today, but the rapid increase in greenhouse gas emissions is resulting in rapid temperature increases that more than offset the cooling effect of aerosols. This is discussed in more detail in the next chapter.

Not only are the daily maximum near-surface temperatures increasing, but the night-time daily **minimum** temperatures are increasing even more (Braganza et al., 2004). Between 1950 and 1993 the average minimum air temperatures over land increased almost 0.2 °C per decade. This is twice as much (0.1 °C per decade) as the daytime daily maximum temperatures over the same period. As a result there has been a lengthening of the freeze-free season in many mid- and high-latitude regions. In western and central North America, the lowest minimum temperatures have risen even more. For the months of January through March the daily minimum air temperature has warmed at rates exceeding 0.3 °C per decade (Robeson, 2004). Since 1980 in the United States, the length of the frost-free season has increased by about one week. The national average increase in the length of the frost-free season from the beginning to the end of the 20th century is about two weeks (Kunkel et al., 2004). Measurements at 246 sites distributed across Canada show that the ground has warmed about 0.7 °C in the last 100 years, and that most of the warming has occurred in the southern part of the country. Unlike the present warming, the signal of the Little Ice Age is not homogeneous across Canada (Beltrami et al., 2003). Temperature measurements from 73 climate stations in New England and New York State over the past 100 years show an average temperature increase of 1.11 °C (Trombulak and Wolfson, 2004). This is twice the Northern Hemisphere increase of 0.6 °C over the same time period.

Temperatures in the atmosphere's lowest 8 km (the troposphere) have also risen during the past 40 years. Until recently, satellite measurements of atmospheric temperatures derived from microwave measurements between 1979 and 2001 have been in conflict with the surface temperature measurements. They appeared to show little or no increase in temperature during this time interval. However, there has been a long-standing problem caused by ongoing cooling of the stratosphere when interpreting satellite measurements of microwave emissions. The earlier reports failed to take into account the fact that the microwave–emitting layer in question extends into the lower stratosphere where strong cooling is occurring due largely to the loss of ozone. When the measurements are adjusted for this cooling of the stratosphere, the satellite temperatures show almost the same warming of the lower atmosphere as the thermometers show at the surface; 0.18 °C per decade, versus the surface measurements of 0.17 °C per decade. The satellite and surface measurements are now in agreement (Prabhakara et al., 2000; Fu et al., 2004). The middle mesosphere of the atmosphere from about 50 to 80 km is cooling at a rate of about 2 to 7 °C per decade, probably as a result of the increasing concentration of greenhouse gases (Beig et al., 2003; also see Chapter 2, Figure 2.6).

Other Indicators of Global Warming

There are other strong indicators of rising global temperatures. Since the late 1960s there has been about a 10% decrease in the extent of snow cover. A reduction of about two weeks in the annual duration of lake and river ice cover has occurred at mid to high latitudes over the 20th century in the Northern Hemisphere. There has been widespread and accelerated melting of mountain glaciers, and now the Greenland ice sheet is beginning to melt (see Chapter 10). The extent of sea ice in spring and summer in the Northern Hemisphere has decreased by about 18% since the 1950s, and the sea–ice thickness has declined by about 40% during the late summer and early fall. Even the extent of winter sea ice has decreased. If we had no other information except the melting ice, we would be forced to conclude that the Earth was warming.

Tide gauge data show that the average global sea level rose about 0.2

meters during the 20th century due to glacial melting and expansion of the water caused by the oceans' temperature increase. The average ocean near-surface temperature has risen 1 °C during the past 50 years, but some areas have risen even more and at a faster rate. For example, the Bering Sea depth-averaged temperature was 2 °C warmer in 2003 than the mean temperature between 1995 and 1997. All of these indications, and more, of a rapidly warming Earth will be discussed in more detail in later chapters. They also have serious consequences for our future, and even now the consequences are being felt with increasing frequency.

NON-UNIFORM TEMPERATURE RISE

The Earth is not warming uniformly. Most of the continental area is situated in the Northern Hemisphere, while most of the ocean area is in the Southern Hemisphere. Consequently, the cooling effects of the ocean are less in the Northern Hemisphere than in the Southern Hemisphere. The northern high latitudes (the Arctic region) are warming faster than anywhere else on Earth.

The average temperatures across the Arctic have increased twice as much as the global average, and some areas have experienced much greater increases. In some areas of Alaska and Canada the temperatures have increased 3 to 4 °C during the past 60 years compared to the global average of 0.5 °C. Thirty years ago in Fairbanks, Alaska, the temperature reached 27 °C (80° F) for a total of only one week. Today it reaches or exceeds that temperature for a total of about three weeks. Furthermore, the ground temperatures are generally increasing at an alarming rate (see Chapter 10). For example, ground surface temperatures in Canada have warmed about 0.7 °C in the last 100 years with the largest warming in southern areas of the country (Beltrami et al., 2003).

One of the worst places on Earth to have warming is the Arctic. This region exerts a strong control over the weather patterns in the Northern Hemisphere, and has a strong influence on the world's climate. Heating of the Arctic will modulate the strength of storms and westerly winds below the Artic, and strongly influence the European climate, for example (Dethloff et al., 2006). The polar regions control the heat balance because more heat is absorbed in the tropics than at the poles. As a consequence, weather systems and ocean currents are constantly transporting heat poleward. In the polar regions extensive ice cover mostly reflects the Sun's radiation back to space, keeping the region cool and a good sink for the heat brought in from lower latitudes. However, if the ice cover melts, as is currently the case, less radiation is reflected back to space decreasing the heat repository and warming the entire planet.

Not all parts of the Earth have become warmer. There is one area that has cooled in the last 30 years: the central part of Antarctica has cooled about 1 °C. The reason for this cooling is apparently the ozone hole in the stratosphere that has concentrated cold air over central Antarctica. Low ozone levels and increasing greenhouse gases in the stratosphere lead to a positive phase of a shifting climate pattern in the Southern Hemisphere called the Southern Annular Mode. This positive phase causes colder air to be concentrated in the central Antarctic interior. Ironically, the increase in ozone levels in coming decades due to the banning of ozone-depleting chemicals by the Montreal Protocol will probably reverse this trend so that central Antarctica will begin to warm up in a manner similar to that on the Antarctic Peninsula and west Antarctica (Shindell and Schmidt, 2004).

West Antarctica and the Antarctic Peninsula have warmed during the same length of time as the cooling of the Antarctic interior. The Antarctic Peninsula has warmed about 2 °C and west Antarctica about 1.5 °C over the past 40 years. This warming has resulted in melting portions of the Antarctic ice sheet, and the collapse of large parts of the Antarctic Peninsula ice shelf (see Chapter 10).

January 2004 Surface Temperature Anomaly (deg. C)

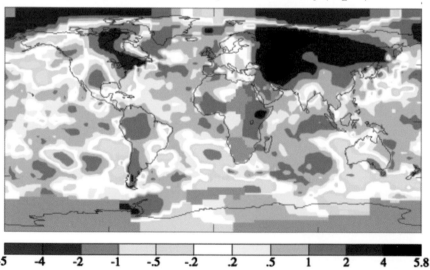

FIGURE 6.7 *This temperature anomaly map for January 2004 shows that the northeast coast of the United States and Canada were experiencing abnormally low temperatures (blue area) while the rest of the world was experiencing abnormally high temperatures. See text for discussion. (Courtesy of the Goddard Institute for Space Studies, New York, NY)*

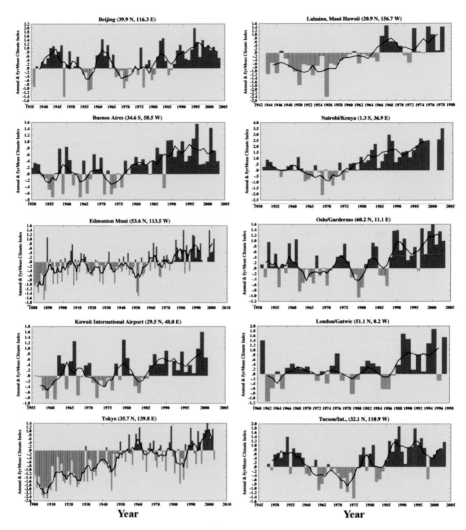

FIGURE 6.8 This shows for various cities the "common sense" index for sensing global warming if a person has lived there most of his life. Red color is a positive indicator for global warming and blue color indicates a negative one. See text for explanation. Most long residents of these cities should now be sensing that the climate is warming. (Courtesy of the Goddard Institute for Space Studies, New York, NY)

"Global Warming, Huh!"

This was the headline in a New York tabloid in January 2004 during an unusual cold spell that was, in fact, no colder than other cold spells in the last century. It is hard for people to take global warming seriously when they are suffering through a cold winter. However, this was nothing more than a natural variation in the weather that produces cold or hot spells.

World wide the Earth was experiencing one of the hottest Januarys on record (0.54 °C above the global average). Figure 6.7 shows the global distribution of temperature anomalies for January 2004. Although the east coast of the United States and Canada was experiencing unusually cold temperatures, the rest of the world was experiencing abnormally warm temperatures. The world does not revolve around the east coast of the United States. When you consider global warming you must look at it from a global perspective or you will reach very wrong conclusions.

James Hansen and his colleagues at the Goddard Institute for Space Studies have devised a Common Sense Climate Index based on climate indicators that people tend to feel. These include such things as seasonal mean temperatures, number of days requiring heating or cooling, frequency of extreme hot or cold temperatures, and record daily high and low temperatures. This index attempts to quantify the degree to which a person can notice climate change. If the index consistently shows a value of 1 or more, then the climate change should be noticeable to most people who have lived at that location for a few decades. Figure 6.8 shows the index for 10 cities in various parts of the world. Currently all have a climate index at or near 1. In the near future global warming should be readily apparent to most people, if it is not already.

SUMMARY

The vast preponderance of scientific studies overwhelmingly indicate that the Earth is rapidly warming. Although there have been other climate changes in the past 10,000 years, the present global warming is unprecedented in its rapid increase and worldwide scope. Furthermore, the warming is not uniform. The high northern latitudes are warming faster than elsewhere. The average global temperature is 0.5 °C higher than the 1951–1980 average. In the Northern Hemisphere it is 0.6 °C higher and in the Southern hemisphere it is 0.4 °C higher. The main cause of this warming is discussed in the next chapter. It is almost certainly due to the rapid rise of greenhouse gases caused predominantly by the human burning of fossil fuels.

7

THE HUMAN
IMPACT

GLOBAL warming is a reality and its causes are now well known. Of the causes of climate change discussed in Chapter 4, only two can operate on time scales short enough to account for today's rapid temperature rise: (1) increase of greenhouse gases and (2) increase in solar irradiance. Although large volcanic eruptions work on a very short time scale, they cool the climate, not warm it. They can warm the climate if there are numerous gigantic eruptions that emit little ash and a lot of CO_2. However, that is not happening now. A rapid change in the present ocean thermohaline circulation would also cool the climate in parts of the high-latitude Northern Hemisphere. Large asteroid or comet impacts can also change the climate in a very short time, but that is obviously not happening today. The other causes, such as continental drift or orbital variations, operate on too long a time scale to account for the current rapid increase in temperature.

INCREASE IN SOLAR IRRADIANCE

The importance of variations of solar irradiance on climate change is still not clearly understood. The problem is that we do not know how much the Sun varies in its energy output over time. One way to attack this problem is to study Sun-like stars to understand their variations in irradiance. We know that the irradiance of the Sun over an 11-year sunspot cycle is only about 0.08% (Foukal et al., 2004). This is almost certainly too little to influence climate in any meaningful way. Most models for the influence of solar irradiance on climate call for the sunspot cycle not only to be superimposed on irradiance variations of at least a similar or greater magnitude, but to take place over periods longer than 11 years. The apparent identification of a solar cause in past climate changes assumes the existence of such variations over longer periods. The models include both the sunspot cycle and a more speculative long-term component. The strength of this second component is about five times that of the sunspot cycle based on luminosity variations of supposedly Sun-like stars. However, this evidence has been weakened by the discovery that the stars in these studies may not be truly Sun-like (Foukal et al., 2004).

Recent studies of truly Sun-like stars, such as 18 Scorpii (HR6060), show no luminosity variations greater than 0.05%. This suggests that the Sun's irradiance variation of 0.08% is no lower than similar stars. If this were the case then variations in the Sun's irradiance would have only very minor effects on the Earth's climate.

Continuous monitoring of solar irradiance for over 20 years indicates that the total amount of radiative forcing has been about 0.12 W/m^2, which is far too little to account for the current rapid rise in the global average temperature. It accounts for only about 3% of the current warming (see Figure 7.4). The main cause of present global warming must be the accumulation of greenhouse gases.

GREENHOUSE GASES AND THEIR RAPID INCREASE

There is now overwhelming evidence that the rapid increase in greenhouse gases is the primary cause of global warming. Recent atmospheric measurements and bubbles of the past atmosphere trapped in ice cores show that the greenhouse gas content increased dramatically during the past 200 years. They are increasing 30 times faster than during the last great Hot House warming 55 million years ago (see Chapter 5). In more recent times we have been adding man-made chemicals to the atmosphere, some of which are truly horrendous greenhouse gases (22,000 times more powerful than CO_2).

There are a variety of greenhouse gases but only four are produced by both nature and humans: carbon dioxide (CO_2), methane (CH_4), nitrous oxide (N_2O), and ozone (O_3). There are a number of other man-made chemicals called halocarbons that are also being emitted into the atmosphere. However, CO_2 is by far the most abundant and is the primary cause of the present global warming. Often in the literature the emissions will be referred to as just carbon, because this is the atom primarily responsible for the greenhouse effect, and it is also a constituent of methane. The equivalent amount of CO_2 for a given amount of carbon is 3.667 times the amount of carbon. Most of the values used here are for CO_2 or CO_2 equivalent. Table 7.1 lists some of the most common greenhouse gases, their lifetimes in the atmosphere, and their greenhouse warming ability relative to CO_2. For example, methane has a lifetime in the atmosphere of 12 years, and over a period of 100 years it would take 23 kg of carbon dioxide to have the same warming effect as 1 kg of methane.

The quantity of greenhouse gases is so small compared to the other constituents of the atmosphere (nitrogen, oxygen and argon) that they are measured in parts per million (ppm), or parts per billion (ppb) or, in some cases, parts per trillion (ppt). For instance, the current amount of CO_2 in the atmosphere is 383 ppm, or 383 parts of CO_2 for each million parts of air. As it is measured by volume of air, it is often written as ppmv, where "v" stands for volume.

Carbon Dioxide

Carbon dioxide (CO_2) is produced naturally by vegetation, decaying organisms, forest fires, exhaled by animals, and volcanic eruptions. It is also produced by humans burning fossil fuels, making cement, and land use. Most carbon dioxide is taken up by forests, grasslands, and oceans. Land plants take up the equivalent of about 220 billion metric tons of CO_2 each year (33.3%), while oceans take up about 330 billion metric tons each year (66.7%). Currently the terrestrial biosphere sequesters about 20 to 30% of global human CO_2 emissions (Gurney et al., 2002; Sarmiento and Gruber, 2002). Hansen and Sato (2004) have estimated that about 42% of CO_2 emitted by human activity is absorbed by the oceans and land.

TABLE 7.1 Estimated global warming potentials relative to carbon dioxide
(Warming effect = kilograms of gas per kilogram of CO_2)

Gas	Lifetime (years)	Warming effect for durations of	
		20 years	100 years
Carbon dioxide	5 to 200[a]	1	1
Methane	12	62	23
Nitrous oxide	114	275	296
Some halocarbons (produced only by humans)			
HFC-23	260	9,400	12,000
HFC-125	29	5,900	3,400
HFC-134a	13.8	3,300	1,300
CF_4	50,000	3,900	5,700
C_2F_6	10,000	8,000	11,900
SF_6	3,200	22,200	32,400

[a] No single lifetime can be defined for CO_2 due to different rates of uptake by different removal processes.
Note: The uncertainty in the global warming potentials is typically about 35%.
Source: UN Intergovernmental Panel on Climate Change.

Of all the greenhouse gases, CO_2 is by far the greatest contributor to global warming (Hansen and Sato, 2004). Its use is increasing very rapidly, and to date little is being done to reduce human-caused emissions. Table 7.2 lists for the periods 1980–1989 and 1990–1999 the CO_2 budgets based on measurements of atmospheric CO_2 and oxygen (O_2), and the estimated ocean emission of O_2. The uptake of CO_2 by the ocean and land is decreasing (Joos et al., 2003) while the amount emitted from land use is increasing. There has been a 29% decrease in ocean uptake and a 43% decrease in land uptake between the 1980–1989 and 1990–1999 periods. Although the fraction of CO_2 uptake by the oceans has decreased,

TABLE 7.2 Global CO_2 budgets (billion metric tons/year)

Process	1980–1989	1990–1999
Atmospheric increase	12.1 ± 0.37	11.73 ± 0.37
Fossil fuel emissions increase	19.8 ± 1.1	23.1 ± 1.47
Ocean–atmosphere uptake	−6.23 ± 2.2	−8.8 ± 2.57
Land–atmosphere uptake	−1.47 ± 2.57	−2.57 ± 2.93
Land use change	7.33 ± 2.93	8.07 ± 2.93

Source: Joos et al. (2003)

the absolute amount has increased since the 1980s because the yearly emissions have been increasing.

The uptake of human-produced CO_2 is strongest in regions of "old" upwelling cold water that has spent many years in the ocean's interior since its last contact with the surface.

Computer models suggest that in the future there will be even more decreases in carbon sinks (Winguth et al., 2005). Future increases in oceanic equatorial upwelling will enhance the outgasing of CO_2 from oceans causing the uptake of CO_2 to decrease by about 16 to 22%, and increases in soil temperatures will reduce the human-caused CO_2 uptake from CO_2 fertilization up to 43%. Therefore, both the land and marine carbon cycle will eventually have a positive feedback on the Earth's climate.

It has been suggested that the increase in CO_2 will be at least partly offset by what is termed "CO_2 fertilization." The concept is that elevated levels of CO_2 would stimulate plant growth so that plants would take up excess CO_2 to produce carbohydrates, which are their stored energy source. However, contrary to predictions, increased CO_2 only accelerates planet growth to about one-third of what was expected. In fact, increased CO_2 may have a positive feedback in that CO_2 is absorbed less with increasing CO_2 levels (Young et al., 2006). The stomata of leaves are the parts of a plant that "breathe" in CO_2 and "exhale" oxygen. A new study shows that the level of CO_2 in the atmosphere controls the opening and closing of leaf stomata (Young et al., 2006); the higher the concentration of CO_2, the smaller the stomata opening and the less CO_2 intake. The lower the CO_2 abundance, the larger the opening and the more CO_2 intake. A doubling of the CO_2 abundance caused leaf stomata to close by about 20–40% in a variety of plant species, thus reducing the CO_2 intake. Therefore, the increasing atmospheric abundance of CO_2 will result in less CO_2 uptake by plants, not more.

Methane

Methane (CH_4) is produced naturally by wetlands, gas hydrates, permafrost, oceans, non-wetland soils, termites, decaying organisms and, surprisingly, vegetation. It was recently discovered (Keppler et al., 2006) that plants give off methane as well as oxygen, and global vegetation may be emitting between 10 and 240 million metric tons each year. Humans produce methane from natural gas mining, biomass burning, landfill, and rice and livestock farming and are responsible for the rapid atmospheric increase of this gas (Ruddiman and Thomson, 2001). It is taken up by microbes, modified by hydroxyl (OH) radicals, and oxidized to CO_2.

Today 30% of the methane comes from natural sources and 70% from human activities. Its greenhouse warming effect is about 62 times that of CO_2 over a 20-year time period and 23 times that of CO_2 over a 100-year period, and has increased 150% over the past 200 years. Methane increased from 1,625 ppb during 1984 to 1,751 ppb during 1999. Since then it has remained nearly constant, which may be the result of about 10 million metric tons of decreased CH_4 emissions above 50 °N latitude in the early 1990s due to a 30% decrease in fossil fuel burning in the former Soviet Union. This may have pushed the global methane abundance toward the present level concentration (Dlugokencky et al., 2003). However, the present level may only be a temporary pause in the increase. Future releases from peatlands and permafrost may eventually reverse this trend.

It was originally believed that methane was responsible for about one-sixth of present-day global warming (IPCC, 2001). However, that figure was based on the current amount of methane in the atmosphere. Because it is converted to CO_2 and other compounds, the contribution to global warming by methane is better estimated by the amount of methane **emissions**, not the current methane content. When this is taken into account, the contribution of methane to global warming is more like one-third (Shindell et al., 2005).

Enormous quantities of methane are locked up as hydrates in permafrost and seafloor deposits (Figure 7.1). Below water depths of about 300 meters the pressure and temperature conditions cause methane to form ice-like crystals called *methane hydrates*. Methane hydrates are white ice-like compounds made up of molecules of methane gas trapped inside cages of frozen water (Kvenvolden, 1999; Suess et al., 1999). One cubic centimeter of ice can have as much as 164 cm^3 of methane—and this ice burns if ignited. It is estimated that about 10 trillion tons of methane are locked up in permafrost and seafloor sediments where it is stable under present temperature and pressure conditions. This enormous amount of methane is twice the amount of organic carbon in fossil fuel reservoirs but the estimate assumes that methane hydrate is always present where the pressure and temperature conditions are suitable. However, this is not always the case so this huge amount may be an overestimate.

In the geologic past, catastrophic releases of methane have probably caused at least one large temperature surge during the Tertiary Period (see Chapter 5). It has also been suggested that catastrophic releases of methane from seafloor hydrates caused some of the abrupt climate changes during the last ice age (Kennett et al., 2003). Observations of the transit of

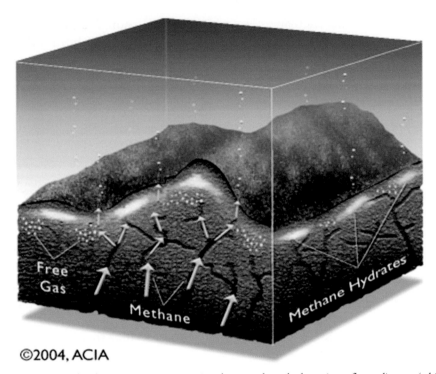

©2004, ACIA

FIGURE 7.1 This diagrammatic representation shows methane hydrates in seafloor sediments (white patches) and methane gas issuing from the sediments. (From the Arctic Climate Impact Assessment Report, 2004)

methane hydrate through seawater show that even small pieces of hydrate (a few centimeters) can survive transit through 800 meters of water to transfer methane directly to the atmosphere (Brewer et al., 2002). It is conceivable that, in the future, global warming could trigger the release of large quantities of methane to abruptly propel us into a Hot House (Leifer et al., 2006). As we will see, global warming is already responsible for the start of methane release from melting permafrost.

Nitrous oxide

Nitrous oxide (N_2O) is produced naturally by a wide variety of biological sources in soil and water, and from human activities by a variety of energy-related, industrial, and waste management activities. For example, nitric acid production for manufacturing fertilizers and fuel combustion produces this gas. It is eventually destroyed in the stratosphere by its reaction with light and excited oxygen atoms over a period of about 115 years. Nitrous oxide has increased about 13% over the past two centuries due to human activity, and it has a greenhouse warming effect about 275 times that of CO_2 over a 20-year time scale. However, CO_2 is by far the most important because there is so much more carbon dioxide compared to N_2O or any other greenhouse gas.

Ozone (O_3)

The greatest abundance of ozone is in the stratosphere where it filters out harmful ultraviolet radiation. Here it does not cause warming because of the very cold temperatures in the stratosphere (See Chapter 2). In the troposphere, however, it causes warming. It is formed by the reaction of sunlight on air pollutants containing hydrocarbons and nitrogen oxides that issue from industrial plants, automobiles and other sources. These compounds react to form ozone directly at the source of the pollution or many kilometers downwind of the pollution source. Ozone is not uniformly distributed in the atmosphere so it is difficult to quantify its potency as a greenhouse gas.

Halocarbons

There are a large variety of these industrial compounds, most of which are extremely potent greenhouse gases (Table 7.1). These gases do not occur in nature. They are man-made and have long residence times in the atmosphere. Halocarbons are strong greenhouse gases that consist of compounds of carbon containing bromine, chlorine, and fluorine. Chlorine-containing halocarbons called *chlorofluorocarbons* (CFCs) destroy the ozone layer in the stratosphere and let harmful ultraviolet radiation penetrate to the surface. The Montreal Protocol treaty that was ratified in 1989 banned CFCs. However, their replacements are synthetic fluorinated compounds (*hydrofluorocarbons* [HFCs] and *perfluorocarbons* [PFCs]) that are not banned by the Montreal Protocol or any other international agreement. These greenhouse gases, together with *sulfur hexafluoride* (SF_6), are just as horrendous as the banned compounds. They are emitted during aluminum production, by the semiconductor and other electronics industries and as refrigerants in air-conditioning units. The primary use of SF_6 is in electric transmission and distribution systems.

One example of these compounds is *fluoroform* (HCF-23, CHF_3) that was discovered in the atmosphere in 1997 (Oram et al., 1998). Fluoroform has one of the greatest warming potentials at 12,000 times that of CO_2 over a 100-year period. In 1997 it had an atmospheric concentration of 14 ppt (parts per trillion) and is increasing at a rate of 5% per year. Its emission rate (in the year 2000) was 7,000 metric tons per year. This may seem like a small amount, but because of its large warming potential, the total amount that has accumulated to date is equivalent to about 2.7 billion metric tons of CO_2. Furthermore, its residence time in the atmosphere is about 260 years, which is longer than CO_2.

The largest emission of any halocarbon is HFC–134a (CF_3CH_2F) at $\sim 25,000$ metric tons per year. It has a warming potential of 1,300 times that of CO_2 over a 100-year lifetime. The total amount of halocarbon emissions is about 360 million metric tons per year and rapidly increasing. These compounds are becoming an important contributor to global warming, currently more effective than N_2O and about the same as ozone.

THE EMISSION OF GREENHOUSE GASES

The present abundance of carbon dioxide, methane and nitrous oxide is greater than any time in at least the past 650,000 years (Spahni et al., 2005; Siegenthaler et al., 2005). Carbon dioxide's current rate of increase is 200 times faster than at any time in the last 650,000 years, and probably considerably longer. There is also a very good correlation between temperature and CO_2 concentration indicating a strong relationship between CO_2 abundance and climate. Similar increases are occurring for methane, nitrous oxide, and halocarbons. Pre-industrial concentrations of methane were 773 ppb during the past 650,000 years compared to about 1,752 ppb today (Spahni et al., 2005). For nitrous oxide the value was about 278 ppb compared to 315 ppb today. Unfortunately CFCs are still being emitted by old technology, the destruction of many old refrigerators, violations of the Montreal Protocol, and other misuses of technology.

Figure 7.2 shows the CO_2, CH_4, N_2O concentrations and a temperature proxy for the past 650,000 years from atmospheric gases trapped in air bubbles in Antarctic ice cores. For CO_2, the concentrations range from minimums of about 190 ppm during glacial periods to maximums of about 280 ppm during interglacial periods. The difference in CO_2 concentrations between glacial and interglacial periods is not well understood. A variety of explanations have been offered for these 90-ppm variations, including variations in the nutrients and alkalinity of the oceans, increased solubility of CO_2 in colder seawater, decreased air–sea exchange in the southern ocean due to enhanced sea-ice cover, and several others. A recent study of stable isotopes in diatoms (small sea creatures) spanning the past 640,000 years indicates that variations in sea-ice cover,

Ice Core Data Showing Greenhouse Gas Concentrations

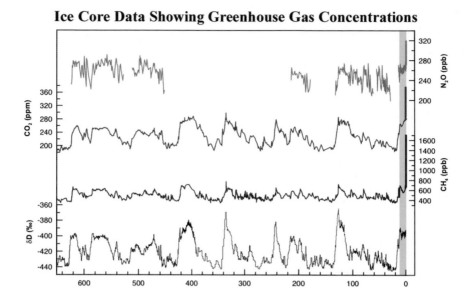

FIGURE 7.2 *Atmospheric carbon dioxide (red), methane (blue), and nitrous oxide (green) concentrations for the past 650,000 years derived from Antarctic ice cores and direct measurements. The bottom black curve is a temperature proxy from the cores. The shaded area is the present interglacial period, the Holocene. The rise in these greenhouse gases is unprecedented during the past 650,000 years. The steep rise began in about 1850 near the beginning of the Industrial Revolution. See text for discussion. (From IPCC 4th Assessment 2007)*

water column stratification, and iron addition from deep water mixing are the most likely causes of the CO_2 variations (Schneider–Mor et al., 2005).

The current large concentration of CO_2 occurred during the last 200 years from about 1800 when it was 280 ppm to the present at 383 ppm. That represents a 35% increase, all of which occurred within the past 200 years.

Figure 7.3 shows the increase in carbon dioxide, methane, and nitrous oxide from 1000 AD to 2000 AD. Also shown are the SO_2 emissions from Europe and the United States, and the sulfate aerosol deposits in the Greenland ice sheet from 1400 AD to 2000 AD. These aerosols cool the atmosphere by reflecting the incoming solar radiation and indirectly by increasing cloud reflectivity. The rapid increase cooled the Earth and caused the relatively flat portion of the temperature curve in Figure 6.5a. The decrease was due to the Clean Air Act in the 1970s that mandated "scrubbers" on coal-fired power plants and reduced aerosol emissions. The upward trend in the greenhouse gas emissions begins about 200 years ago near the start of the Industrial Revolution. Since that time there has been a 35% increase in carbon dioxide, a 150% increase in methane, and a

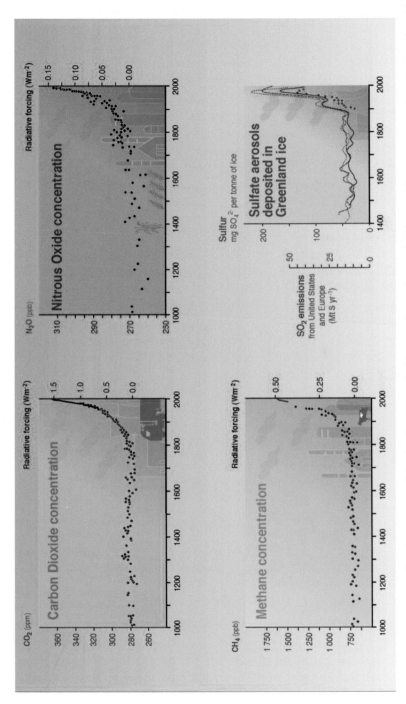

FIGURE 7.3 These diagrams show the increase in CO_2, N_2O, CH_4 and sulfate aerosols over the past 1,000 years. The increases start about 1850 or about 100 years after the start of the Industrial Revolution. They also begin a rapid increase at the same time as the beginning of the rapid increase in the world population. This implies a human source. Aerosols cool the atmosphere and there was a dramatic decrease beginning in the 1970s when coal plant scrubbers were mandated. (From the UN IPCC report, 2001)

17% increase in nitrous oxide. Also during that time there has been a 35% increase in tropospheric (lower atmosphere) ozone resulting in a radiative forcing of about 0.35 W/m^2. This is strong evidence that humans are the cause of the rapid rise. The radiative forcings for the CO_2, N_2O, CH_4 and halocarbon concentrations have been estimated at 1.6, 0.16, 0.5, and 0.34 W/m^2 respectively. The total greenhouse gas forcing is about 3.0 W/m^2, enough to account for the present global warming. Figure 7.4 shows the radiative forcing of the main factors that effect the global average temperature. Of the greenhouse gases, the most powerful is CO_2 that accounts for about 56% of the warming because it is the most abundant. This is followed by methane at 16%, ozone at 12%, halocarbons at 11% and nitrous oxide at 5%. Increased solar irradiance accounts for only about 3% of the total warming.

Aerosols cause cooling in two ways; a direct effect and an indirect effect. In the direct effect, aerosols directly reflect sunlight back to space cooling the Earth. Clouds only form with the aid of aerosols that form nuclei around which water condenses. Sulfate aerosols form in the atmosphere as a result of the release of SO_2 from burning fossil fuels. The SO_2 is subsequently transformed to sulfate aerosols (SO_4). These enhance the ability of the particles to serve as cloud condensation nuclei leading to a general decrease in cloud droplet size resulting in a larger surface area of the droplets. This, in turn, enhances the reflectivity of the clouds causing more sunlight to be reflected back to space. Also the generally smaller cloud droplets mean that coalescence will be suppressed making the clouds more persistent. Therefore, both their coverage and average water content increase, again leading to enhanced cloud reflectivity.

It is estimated that the aerosol direct effect causes a cooling of about $-0.5\ Wm^{-2}$, and the indirect effect a cooling of about $-0.7\ Wm^{-2}$ shown in Figure 7.4. This is a combined cooling of about $-1.2\ Wm^{-2}$. If the aerosols where not present then the current global average temperature anomaly would be about 1.4° C instead of 0.5° C. A recent study (2007) by NASA's Goddard Institute for Space Studies shows a gradual decline in aerosols since 1990. By 2005, global aerosols had dropped as much as 20% from the relatively stable level between 1986 and 1991. If they keep dropping, then warming will accelerate even faster.

The Keeling Curve

This CO_2 abundance curve is named after the late scientist who was first to measure carbon dioxide in the atmosphere on a continuous basis. Under his aegis the Mauna Loa Observatory on the island of Hawaii was established in 1958 in the middle of the Pacific Ocean away from

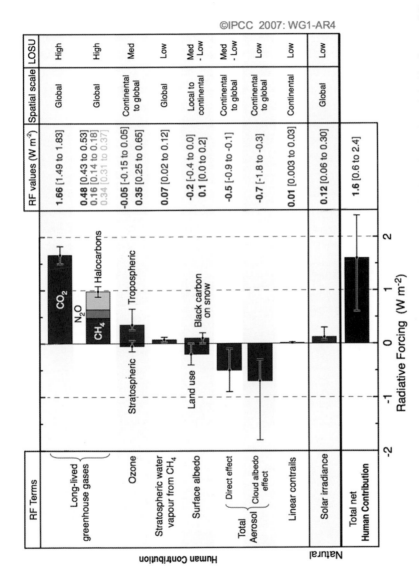

FIGURE 7.4 Radiative forcings due to various causes are shown in this diagram. The red forcings (right) heat the Earth, and the blue forcings (left) cool the Earth. Greenhouse gases are mainly responsible for the current warming and far outweigh the others. However, the human-caused aerosols have produced a significant cooling which has kept the temperature from reaching even higher levels than today. (Modified from UN IPCC 4th Assessment 2007)

pollution sources. In this way the atmosphere's composition could be monitored without fear of contamination from nearby industrial sources. Since 1970 over 60 other stations around the world have been monitoring atmospheric greenhouse gas abundances, but the Mauna Loa station has the longest record.

I was visiting an astronomical observing test site on Mauna Kea volcano and doing some geological fieldwork on Mauna Loa in 1964. When I visited the Observatory on Mauna Loa, one of the meteorologists showed me a chart documenting the dramatic rise in CO_2 over the five years since the station began to record complete annual measurements in 1959. He was not sure what was causing the rise, and when I asked if he was going to publish the results of the measurements, he said that it might be just a fluctuation that would return to "normal" in a few years. (At that time there were no ice core data or climate proxies to tell us about past concentrations.) That was over 40 years ago and the CO_2 content has continued its rapid rise with no end in sight.

Figure 7.5 shows the Mauna Loa measurements of CO_2 from 1959 to 2007. Since measurements began in 1959, the CO_2 content has steadily

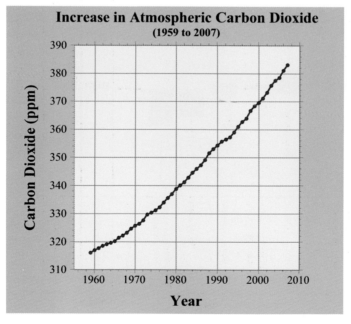

Figure 7.5 The rapid rise in atmospheric CO_2 from direct systematic measurements that begun on the Island of Hawaii in 1958 is shown on this graph. The rise shown in this graph is the mean annual CO_2 abundance, and is called the Keeling Curve. The annual measurements for a given year are from March to March because of the natural seasonal variation in CO_2 abundance. For example, for the year 2000 the measurements are from March 1999 to March 2000. (Data from the Mauna Loa Observatory)

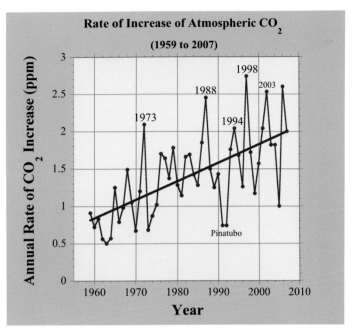

FIGURE 7.6 *The variation of the yearly rate of CO_2 increase is shown on this graph. Also shown are El Niño years that coincide with particularly rapid rises in CO_2. Not indicated is the relatively weak El Niño that began in the Spring of 2006 and ended in early 2007. The red line is a linear fit to the data. There has been a rise in the average rate from about 0.8 ppm/yr in 1960 to a present average rate of 2 ppm/yr today. (Data from the Mauna Loa Observatory)*

increased from about 316 ppm to 383 ppm in 2007. In just the 15 years between 1990 and 2005, the radiative forcing of greenhouse gases has risen over 21% (Hofmann et al., 2006). The rate at which this increase has occurred has not been constant. The rate of increase has been rising from an average of about 0.9 ppm per year to its current average rate of 2 ppm per year (see Figure 7.6). There have been four sharp increases in CO_2 content, but these occurred in El Niño years when the abnormally warm Pacific Ocean waters release CO_2 into the atmosphere. They occurred in 1973, 1988, 1994, and 1998 (shown in Figure 7.6). The amount of CO_2 increase in the first three El Niños was about 2 ppm. However, in the 1998 El Niño the increase was 2.74 ppm. This abnormally large increase was probably due to a strong El Niño combined with the human-caused release of the equivalent of 3 to 9.4 billion metric tons of CO_2 from the burning of peat and forests during the massive 1997 fires in Indonesia (Page et al., 2002).

In 2002 and 2003 there were two large increases with a combined rate increase of about 2.5 ppm. The year 2003 was a relatively weak El Niño year and certainly contributed to the rapid rise, but it could not account

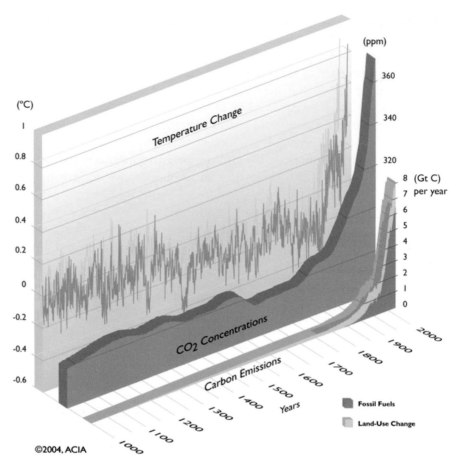

FIGURE 7.7 *This graph shows the relationship between the temperature change, the change in atmospheric CO_2 concentration, and the change in human-caused carbon emissions. All three increase sharply at the same time indicating human-caused greenhouse gas emissions is the main cause of global warming. (From Arctic Climate Impact Assessment, Impact of a Warming Arctic: Arctic Climate Impact Assessment, 2005)*

for the entire large rate of increase in 2002 and 2003. In 2004 the rate of increase returned to about 1.8 ppm and it returned to about 1 ppm in 2005. During 2002–2003 there was no unusual increase in CO_2 from human sources. In fact, the human CO_2 production was down somewhat from previous highs because the world was coming out of a recession. Furthermore, there was no unusual change in the increase in CO_2 at the Southern Hemisphere stations, so the change must have been confined to the Northern Hemisphere. One possible contributor to the large increase may have been drought conditions that were particularly severe from 1999 to 2003. Drought conditions result in a net loss of carbon from land.

However, an unusually sharp increase of 2.6 ppm occurred between 2005 and 2006 when there was no El Niño or unusually large increase in human-caused CO_2 emissions. The reason for this increase is not clear. There was, however, a relatively weak El Niño between the Spring of 2006 and early 2007 that may account, at least in part, for the 2 ppm increase shown for the year 2007 in Figure 7.6. We will just have to keep monitoring the increases to ascertain whether or not a major natural release of CO_2 is occurring.

The rate of atmospheric CO_2 and temperature increases has been the same since the mid-1970s when the atmospheric temperature began its steady rapid rise. Also the rate of human-caused CO_2 emissions and temperature rise is also the same, again indicating that humans are the cause of global warming. Figure 7.7 shows virtually the same rate of increase of temperature, atmospheric CO_2 concentrations, and human-caused emissions of CO_2.

HUMAN-RELATED PRODUCTION OF GREENHOUSE GASES

By far the greatest contributor to global warming is carbon dioxide. This is followed by methane and ozone, which are also important contributors to global warming. The contribution of halocarbons is now about the same as ozone and rapidly growing. As shown in Figure 7.8, the yearly world emission of carbon dioxide is about 25 billion metric tons from the burning of fossil fuels (oil, coal, and natural gas) and cement production. There has been an almost exponential increase in the emission of human-caused CO_2, with a particularly steep rise after World War II.

Human land use also releases CO_2. Figure 7.9 shows the damage done to the Amazon rainforest by humans and human-induced climate change. This is happening to an even greater extent to other rainforests, particularly in Asia. There are 2 million km^2 of tropical forests on the Earth. About 292,000 km^2 of tropical forests that took up CO_2 were destroyed between 1980 and 1995. Today that destruction continues at a net loss rate of 73,000 km^2 per year (UN Food and Agriculture Organization, Nov. 2005). This figure takes into account forest growth from new planting and natural expansion of existing forests. When forests are harvested, burned, or cleared by humans, or there are natural events such as fire, some of the stored carbon is released into the atmosphere. The clearing and degradation of forests accounts for about 20% of the annual

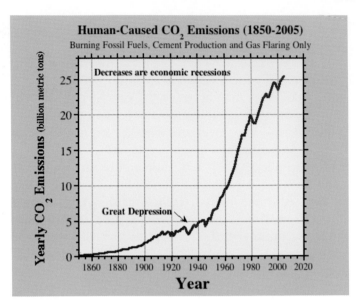

FIGURE 7.8 *This graph shows the worldwide human-caused yearly emission of CO_2 from 1850 (100 years after the start of the Industrial Revolution) to the present. The rapid rise beginning about 1950 coincides with the rapid increase in atmospheric CO_2 and the rapid increase in the world population. (Source: US Department of Energy)*

worldwide emissions of CO_2. The amount of CO_2 released from land use is much more uncertain than that from burning fossil fuels and cement production. It is estimated that about 6.2 ± 4.5 billion metric tons of CO_2 equivalent are released each year by various human uses of the land (Houghton and Skole, 2000). Figure 7.10 shows the estimated increase in the release of CO_2 equivalent from 1850 to 2000. Forest clearing causes cooling because the cleared area is more reflective than a forested area. Worldwide this results in a radiative forcing of about -0.2 Wm^{-2}. However, this is more than offset by the loss of the CO_2 uptake and the increased emission of CO_2 by burning the trees.

The release of CO_2 to the atmosphere by natural and human-caused burning of continental biomass (forests, savanna, etc.) is significant. It is estimated that, in the beginning of the 20th century, between 5.5 and 9.9 billion metric tons of CO_2 equivalent were released each year to the atmosphere from biomass burning, and by the end of the 20th century the release rate was between 5.5 and 12.2 billion tons (Mouillot et al., 2006).

The world's soils hold about 1,500 billion metric tons of organic carbon. Recent studies have shown that as the Earth warms, soil will begin releasing carbon to the atmosphere causing a positive feedback (Knorr et al., 2005; Heath et al., 2005; Schulze and Freibauer, 2005). In the long

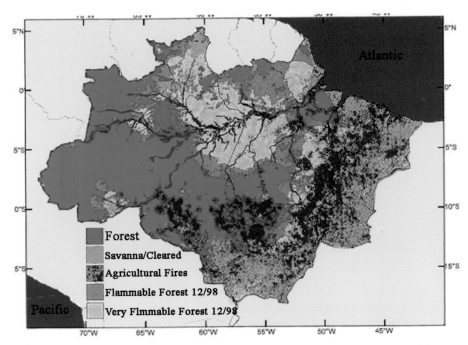

FIGURE 7.9. *This map shows the damage done to the Amazon rainforest by humans and human-induced climate change. The green areas are healthy rainforest while the other areas have been either cleared, burned, or are stressed due to climate change drought conditions. Rainforests in other parts of the world are in even worse condition. (From Nepstad et al., Forest Ecology & Management, 154, 2001)*

term (~ 50–75 years), rising temperatures will speed up the decomposition of organic carbon in soils, which, in turn, will release CO_2 into the atmosphere in excess of any carbon trapped in the soil.

The emission of CO_2 has accelerated almost exponentially during the last 100 years due to industrial growth and the unrestrained growth in the human population (see Chapter 9). Currently the amount of CO_2 released to the atmosphere is about 25 billion tons from burning fossil fuels and cement production, and about 6 billion tons of CO_2 equivalent from land use or a total of 31 billion tons. About 42% of that total is absorbed by the Earth, leaving 18 billion tons in the atmosphere. For human-produced methane, about 2 billion tons of CO_2 equivalent is retained after absorption or conversion, and for human-produced N_2O about 3 billion tons of CO_2 equivalent is being retained. Therefore, we are adding roughly 23 ($\sim \pm 5$) billion tons of CO_2 equivalent to the atmosphere each year.

During the past 10,000 years the Earth absorbed each year most of the greenhouse gases emitted naturally. This partial equilibrium condition kept the greenhouse gases at a more-or-less constant level of about 280 ppm

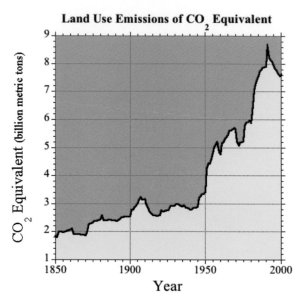

FIGURE 7.10 Estimated land use emissions of CO_2 equivalent from 1850 to 2000. (Data from Houghton and Skole, 2000)

in the case of CO_2. At the start of the Industrial Revolution in 1751 things began to change rapidly. As industry developed and the population grew, more and more greenhouse gases were emitted until the Earth could no longer absorb them all and they began to accumulate in the atmosphere, slowly at first and then faster and faster as industrial production increased and the world population grew. We first began to lose equilibrium about 1850. Now we are so far out of equilibrium that the rate of increase is very rapid. Even worse, the CO_2 sinks are beginning to decrease because of global warming, which will add to the problem. Currently, we would have to reduce our world greenhouse gas emissions by about 60% in order to stabilize the atmosphere. If only CO_2 is considered, then a reduction of about 70% is required for this gas. The longer we wait, the larger percentage we will have to cut.

HUMAN CAUSE OF GLOBAL WARMING

It should be clear by now that humans are the primary cause of global warming, but there is other evidence. Figure 7.11 shows the results of computer simulations of the rise in global temperature from the effects of both natural forcing and human forcing due to the emission of greenhouse

a

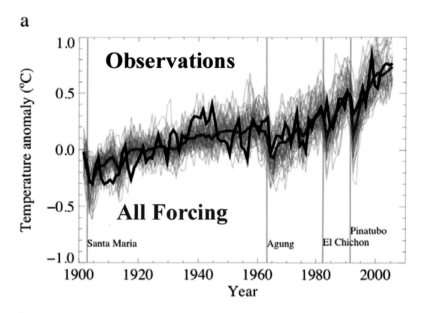

b

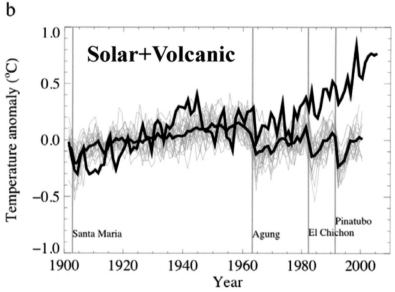

FIGURE 7.11 *This is a comparison of computer simulations of global average temperature anomalies and the observed temperatures (black line) from 1900 to the present (2005). The upper graph shows all forcings (red) including human-caused greenhouse gases, and the lower graph is only natural forcing by solar irradiance and volcanism (blue). They show that only human-caused greenhouse gases can account for the observed warming. The named vertical bars are major volcanic eruptions that lower the temperature. These computer simulations are additional evidence that humans are the primary cause of global warming. See text for explanation. (From the UN IPCC 4th Assessment 2007)*

gases. The black line is the observed temperature trend from 1900 to 2005. In the upper graph the red line is a computer simulation of all forcings including natural and human. In the bottom graph the blue line is only natural forcings including volcanism and solar irradiance. Only the simulation using the human contribution matches the observed temperature curve. This strongly indicates that human–caused emission of greenhouse gases is the primary cause of global warming.

Recent studies of the world's oceans have also shown at a confidence level of more than 95% that human–related emission of greenhouse gases is the primary cause of present global warming (Reichert et al., 2002; Barnett et al., 2005 and 2001; Levitus et al., 2005 and 2001; Fyfe, 2006). There have been large-scale increases in the heat content of the world's oceans during the past 45 years, and this oceanic heat increase is over 10 times greater than the heat increases of the land, atmosphere, or cryosphere (ice)—see Table 6.1 in Chapter 6. State-of-the-art climate models demonstrate that the character of these ocean temperature changes can only be replicated by observed and estimated human emission of greenhouse gases. In all ocean basins, the warming in the upper 700 meters predicted by the emission of greenhouse gases corresponded to the measurements obtained at sea. The agreement was particularly strong in the upper 500 meters. Efforts to explain the ocean changes through naturally occurring variations in the climate or external forces, such as solar irradiance, failed to reproduce the observed temperature increase. The very high degree of agreement together with its statistical significance strongly indicates once again that the warming is the result of human emission of greenhouse gases.

There have been a large number of other recent studies on the causes of global warming based on observed and proxy atmospheric temperature changes (for example, Andronova and Schlesinger, 2000; Keeling and Garcia, 2002; Jones et al., 2003; Bauer et al., 2003; Karoly et al., 2003; Stott, 2003; Andronova et al., 2004; Crowley, 2000; Christidis et al., 2005). All of these studies conclude that the emission of greenhouse gases by human activity is the primary cause of the present global warming. Furthermore, for the first time radiative forcing related to increased greenhouse gas concentrations has been directly measured in Europe (Philipona et al., 2004). The amount of forcing measured is $+1.8 \pm 0.8$ W/m^2, which is about the same as that calculated from the rise in temperature.

There is absolutely no question that the rise in greenhouse gases is the result of human burning of fossil fuels and other human activity. Aside from measuring the human output of greenhouse gases, there is also very compelling evidence from the correlation of the rate of CO_2 increase compared to the rate of human population growth. In Chapter 9, Figures 9.4 and 9.5 show that the rate of increase of atmospheric CO_2 and the rate of increase in human emission of CO_2 is directly correlated with the rate of growth of the human population. There is no doubt that we are the primary cause of present global warming.

FUTURE EMISSIONS AND TEMPERATURES

The future emission of greenhouse gases is hard to predict because it depends on the world economy, world population, efforts to curb emissions, and many other societal conditions. Figure 7.12 shows the possible growth in atmospheric CO_2 concentrations over the next 100 years. They are compared to the CO_2 levels for the past 450,000 years derived from ice core data.

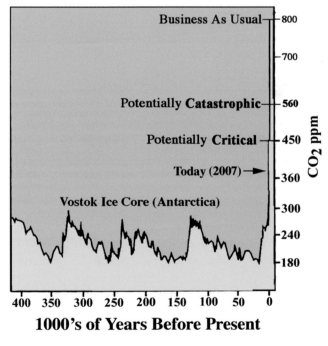

FIGURE 7.12 *This graph shows the possible increase of carbon dioxide compared to their concentrations during the past 450,000 years. For Business As Usual projected emission rates of CO_2 could reach ~ 800 ppm or higher by 2100. The best estimates are CO_2 values between 450 and 560 ppm that are potentially critical to potentially catastrophic. See text for explanation.*

The most optimistic case has emissions of CO_2 peaking at about 36 billion metric tons per year in about 2050, then decreasing to about 18 billion metric tons per year by 2100. This results in an atmospheric CO_2 level of about 560 ppm in 2100. In a more pessimistic case the CO_2 emissions peak at about 110 billion metric tons per year in about 2080, staying constant to 2100. Under this scenario the atmospheric concentration reaches about 800 to 1,000 ppm. As we will see later, either projection has extremely serious consequences.

Projections of future temperatures are based on very complex computer models called Global Circulation Models (GCM). They are run on very fast computers that still take hours or even weeks to produce results. They usually use a doubling of the CO_2 content over pre-industrial levels or about 560 ppm, and run the model for a time period representing ~ 100 years. They also depend on the sensitivity of the climate system to things like feedback processes and forcings, as low sensitivities give lower temperatures than high sensitivities. The models also depend on the number of parameters that can be run simultaneously—the more parameters, the greater the accuracy. Recent studies indicate that the lower value of $1.5\,°C$ increase in 100 years (2001 IPCC estimate) is unrealistically low and that the bare minimum is $2\,°C$ which, although possible, is still less likely than higher temperatures (Stainforth et al., 2005).

A recent study—the largest ever climate-change computer experiment—suggests that we may have dramatically underestimated future global warming (Stainforth et al., 2005). The study used over 45,000 personal computers that were not being used by their owners, multiple parameters running simultaneously, 2,578 simulations, and a run time of many months. In this distributed-computing study they found that perturbed parameters did in fact interact in a non-linear fashion to increase climate sensitivity, raising temperatures greater than expected. After eliminating physically implausible simulations, none of the remaining model versions gave a sensitivity less than $1.9\,°C$. Even this lowest temperature increase was considered unlikely. Figure 7.13 shows possible future temperature increases. With no parameters perturbed, a warming of $3.4\,°C$ occurred. They found that the planet's global temperature could climb to an upper limit of $11\,°C$, and although this is probably unrealistically high, it also appears to be possible. A study that took into account Northern Hemisphere climate data for the past 700 years found that the climate sensitivity is 1.5 to $6.2\,°C$ for a doubling of CO_2 and that there was only a 5% chance of a temperature rise of $11\,°C$ (Hegerl et al., 2006).

For a doubling of CO_2 over pre-industrial levels most modelers think

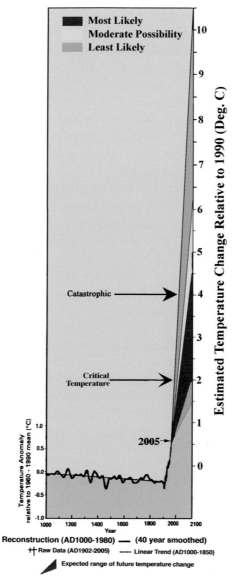

FIGURE 7.13 *This shows the possible increases in Northern Hemisphere temperature over the next 100 years (to year 2100) for a doubling of CO_2 (560 ppm) above the pre-industrial era. The temperatures from 1000 to 1880 are based on multiple proxies. The uncertainty in the projected temperatures is largely the result of various sensitivities of the atmosphere to greenhouse warming. The upper limit is from computer simulations by Stainforth, et al., 2005 and the lower limit is also based on a variety of computer models. Most modelers believe the global average temperature will rise about 3.0 to 4.0 degrees in the next 100 years unless greenhouse gas emissions are curbed. The arrows indicate the estimated global average temperature anomalies potentially critical and potentially catastrophic to civilization. See text for explanation. (Modified from the UN IPCC, 2001 report)*

that a 2 to 4 °C warming is the most likely, and that a 6 to 7 °C warming was possible but much less likely. Figure 7.14 shows a series of computer simulations of future global average temperatures for three scenarios. These simulations depend on certain assumptions and interactions that give different results, none of which are good for humans and other species. To put this in perspective, Figure 7.15 is a diagrammatic representation of the climatic variations during the past 65 million years compared with a 2, 4, and 6 °C global average temperature increase by 2100 AD. The global average temperature anomalies would be 16, 18, and 20 °C for these increases. The temperatures are based on a doubling of the atmospheric CO_2 content over pre-industrial levels by 2100 AD when the most likely increase is considered to be between 2 to 4.5 °C. However, a 2 °C rise is considered critical for civilization and a 4 °C rise is considered catastrophic (see Chapter 12). A global average temperature of 18 °C is within a Hot House, and about half-way to the peak temperatures in the Late Cretaceous Hot House about 65 million years ago.

Atmospheric aerosols and particulates counteract global warming by reflecting incoming radiation back to space. This leads to large uncertainties in the sensitivity of climate to human perturbations, which may lead to large underestimates of future warming (Andreae et al., 2005). These aerosols and particulates are mostly due to pollutants from burning fossil fuels. They appear to have resulted in what has been called "global dimming." However, it is not global in scope; locally the aerosols and particulate matter may have blocked up to 22% of the sunlight from heating the surface (Stanhill and Cohen, 2001), but the global average may be only about 2% at most, and it may have stopped altogether by 1990. Recent NASA results indicate aerosols have decreased about 20% between 1990 and 2005.

Recent studies estimate aerosol radiative forcing anywhere between − 1.0 and −1.9 W/m^2, including cloud-brightness radiative forcing due to aerosols (Bellouin et al., 2005; Kaufman et al., 2005; Rotstayn and Liu, 2005). This means that greenhouse warming has been partly offset by particulate and aerosol cooling (see Figures 7.3 and 6.4). In other words, two of our pollutants have been partially canceling each other out, which suggests that climate may be more sensitive to the greenhouse effect than previously thought. Greenhouse gas levels are projected to rise in the future, but particle and aerosol pollution are now being brought under control, and particles and aerosols settle out much faster than CO_2 can be absorbed (Andreae et al., 2005). Of course, this means that we will have reduced cooling and increased heating at the same time, and, if this happens, then even the more pessimistic global warming forecasts may be

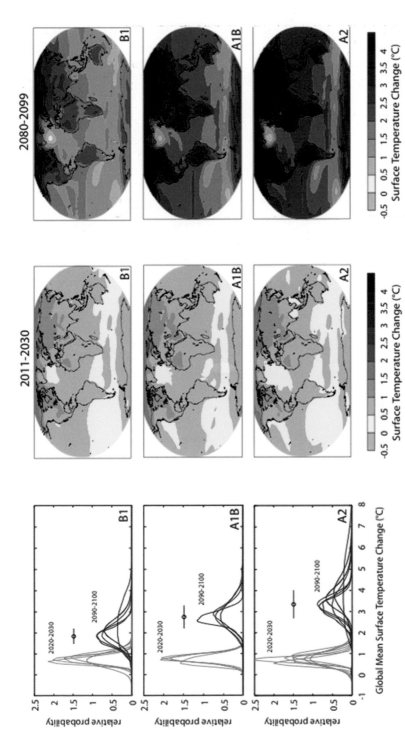

FIGURE 7.14. The projected temperature increase from computer models is shown for three scenarios in this diagram. The upper scenario is most optimistic and the lower produces the highest temperatures. The left side shows the most likely increases from 2020 to 2030 and from 2090 to 2100 for a doubling of CO_2 for different scenarios. The right side shows the distribution of surface temperature increases for the same time periods. Recent studies suggest these computer models may be underestimating the temperature rise. (From the IPCC 4th Assessment Report)

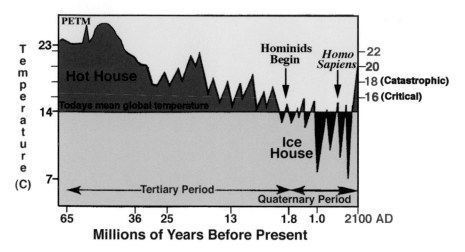

FIGURE 7.15 A quantitative cartoon of the climatic variations during the past 65 million years compared to a 2, 4, and 6 °C global average temperature increase by 2100 AD. The global average temperatures would be 16, 18, and 20 °C for these increases. The temperatures are based on a doubling of the atmospheric CO_2 content over pre-industrial levels by 2100 AD. The most likely increase is considered to be from about 2 to 4 °C. Although a 6 °C increase is possible it is considered much less likely. However, a 2 °C rise is considered critical for civilization and a 4 °C rise is considered catastrophic. Also shown are the approximate times when Hominids and Homo sapiens first emerged and the Paleocene/Eocene Thermal Maximum (PETM). (Modified from Cook et al., 1991. See also Figure 5.2(a))

more likely than not. If there was no aerosol cooling then the global average temperature anomaly would be about 1.4 °C instead of the current 0.5 °C.

Several recent studies have suggested that computer models may have significantly underestimated the temperature rise by the end of the 21st century for a given level of CO_2. Each of these studies approaches the problem in a different way but comes to basically the same conclusion. One uses the ancient climate data from proxies, another uses temperature proxies and CO_2 abundances from ice cores, and yet another uses known temperature changes during the 20th century.

Studies of ancient climates suggest that climate models may have underestimated future warming. In one study it was found that between 84 and 100 million years ago in the Cretaceous "Hot House," surface water temperatures in the tropical Atlantic were 33 to 42 °C and atmospheric CO_2 concentrations were 1,300 to 2,300 ppm (Bice et al., 2006). Computer models that use these CO_2 concentrations do not produce such high water temperatures, nor do even extraordinarily high concentrations of methane produce these kinds of temperatures. This suggests that the model's response to greenhouse gases underestimates the actual climate systems response, possibly by missing some critical factor that amplifies the heating. That critical factor may be the temperature itself.

A similar situation concerns the Paleocene/Eocene warming 55 million years ago (Sluijs et al., 2006). At that time the surface water temperatures in the Arctic Ocean were about 24 °C, and CO_2 levels were about 1,000–1,500 ppm. However, climate models for that period do not come close to simulating such warm waters even with CO_2 levels as high as 2,000 ppm, which indicates that the models have omitted some crucial factor.

Another study using past temperature proxies and the CO_2 abundances from ice core data finds that the increasing temperatures increase the CO_2 abundance by releasing them from natural sources, resulting in a positive feedback (Scheffer et al., 2006). These positive feedback processes can be long term (centuries or longer) or very short term (several years). Scheffer et al. estimate that the actual warming due to burning of fossils fuels may be 15–78% higher than warming estimates that do not take into account the effect of these feedback mechanisms, which in fact most models do not consider. These authors believe that it is probably at least 50% higher because methane and other greenhouse gases are not taken into account. Another study using positive feedbacks also found that temperatures will be higher than predicted by current computer models (Torn and Harte, 2006).

Using known changes in the temperatures during the 20th century, another study finds that computer models have probably underestimated the amount of future warming from a doubling of CO_2 over the pre-industrial level (Shukla et al., 2006). Based on a statistical procedure called "relative entropy," which determines the difference between two probabilities, they find that the upper temperature limit of the computer models is more likely.

If these studies are correct, the temperature increase by the end of this century may be nearer 4.5 to 6 °C rather than 3 °C for a doubling of CO_2 above the pre-industrial level.

Today the atmospheric CO_2 content is 383 ppm and the average global temperature has increased 0.5 °C over the past 30 years. In order to limit warming to 2 °C above pre-industrial levels with a high degree of confidence, the equivalent amount of CO_2 is required to stay below 400 ppm (Schellnhuber et al., 2006). On the other hand, if the concentration of CO_2 reached 540 ppm, then it is unlikely that the global mean temperature would remain under 2 °C. The best estimate is that when the CO_2 content reaches about 440 ppm, the average global temperature will in the future rise to a minimum of 1.5 °C above today's value to 2 °C even if our greenhouse gas emissions went to zero (Knutti et al., 2005). This appears to be an optimistic estimate considering the studies described above. A rise of 2 °C above the pre-industrial value is considered to have serious consequences for humans and other forms of life (O'Neill and Oppenheimer, 2002; Hansen, 2005; International Climate Change Taskforce, 2005; Schellnhuber et al., 2006). The CO_2 content is currently rising at a rate of almost 2 ppm per year, so in about 30 years we will reach 440 ppm unless we cut our emissions drastically before then. This is almost surely a maximum time, because it does not take into account a probable increase in the rate of emissions, a larger temperature rise due to a greater sensitivity of the climate system to greenhouse warming, or an atmospheric reduction in aerosols that cool the climate. In fact, we probably have no more than 15–20 years in which to take serious and vigorous actions to cut greenhouse gas emissions.

We do not have time to spare. We must act now. Delaying action will require a much greater effort later to achieve the same temperature target. Even a 5-year delay is significant, given the current increase in CO_2 emissions. If action is delayed 20 years, rates of emission reduction will need to be 3 to 7 times greater to meet the same temperature target (Schellnhuber et al., 2006). In the absence of urgent and strenuous reduction in greenhouse gas emissions, the world will be committed to at least a 0.5 to 2 °C rise by 2050, and it could be considerably more because of the factors mentioned earlier.

None of the greenhouse gas or temperature projections take into account the possibility of crossing a threshold that leads to an abrupt climate warming by the catastrophic release of natural greenhouse gases or some other cause. Although this is considered unlikely, we do not know in detail how these abrupt changes are triggered. Could the rise of atmospheric greenhouse gases and the complex interactions of other warming conditions set one of these events into motion? We do not know, but if it happened we would be in the worst trouble imaginable.

SUMMARY

Today the greenhouse gas abundance is greater than at any time in the past 650,000 years, and probably a lot longer. Currently we are adding about 23 billion tons of CO_2 equivalent to the atmosphere each year. The present rate of increase is 30 times faster than the emission rate during the great warming event 55 million years ago. Furthermore, certain green-house gas sinks are beginning to decrease which will add to the problem in the future (see Chapter 10). There is no question that the rising greenhouse gas abundance is the primary cause of global warming, and that human activity is responsible for that rise. It is clear that humans are the primary cause, and ultimately, the solution to global warming.

Today (2007) the atmospheric CO_2 content is 383 ppm and rising at an average rate of almost 2 ppm per year. The average global temperature has risen $0.5\,°C$ over the past 30 years and it is estimated that an increase to $2\,°C$ would be critical for humans. An increase to $2\,°C$ is unavoidable when the CO_2 abundance reaches about 440 ppm. We need to reduce greenhouse gas emission by about 60% (about 70% for CO_2 alone) in a short period of time (~ 20 years) to avert a potential calamity.

8

THE MAIN
OFFENDERS

THE emission of greenhouse gases is proportional to a country's population and economy. A very populous country with a modest economy can produce more greenhouse gases than a very wealthy country with a small population. Therefore, the largest emitters of greenhouse gases are the wealthy industrialized nations and the countries with the largest populations. Figure 8.1 shows the emission of CO_2 from only the burning of fossil fuels for various regions of the world in 1993 and 2002. The largest emitter is Asia plus Oceania because several of the largest emitters (China, Japan, and India) are located there. It is followed by North America and western Europe, two highly industrialized and wealthy regions. The emissions have significantly increased from 1993 to 2002 in all regions except eastern Europe plus Russia, where they actually decreased. This is due to the economic decline of the region following the break-up of the old Soviet Union in 1990. In fact, Russia is emitting significantly less CO_2 today than it was in 1990, but it is still the third largest emitter in the world.

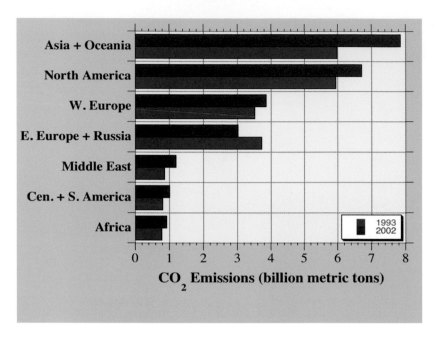

FIGURE 8.1 The chart shows the difference in the yearly emissions of CO_2 for various regions in 1993 and 2002. The decline in emissions between 1993 and 2002 in eastern Europe and Russia is due to the breakup of the old Soviet Union. See text for explanation. (Source: US Dept of Energy)

THE TEN GREATEST EMITTERS OF GREENHOUSE GASES

The 10 countries with the largest emission of CO_2 from energy consumption (burning fossil fuel) are shown in Figure 8.2. These 10 countries account for about 70% of the total world emission of CO_2. In order to stabilize the atmospheric composition of CO_2, we would have to reduce the world emissions of greenhouse gases by about 60%. By far the largest emitter of greenhouse gases is the United States which produces about 25% of all greenhouse gases. Although China and India are not as industrialized as the United States and have a much lower per capita emission output, they produce large amounts of CO_2 (China, 4.7 billion metric tons; India, 1.1 billion metric tons) because they have huge populations (China, 1.3 billion; India, 1.0 billion).

Figure 8.3 shows the yearly per capita output of CO_2. (Although they are not major emitters of CO_2, Ireland and New Zealand would also be

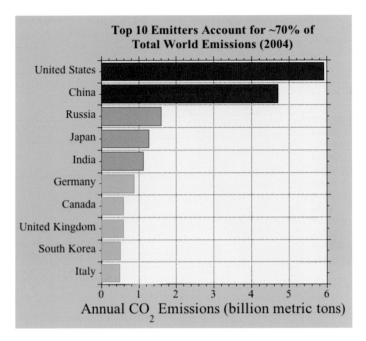

FIGURE 8.2 This bar chart shows the yearly CO_2 emissions for the top 10 emitters. (Source: US Dept of Energy)

high on the per capita list.) Surprisingly, Australia has the highest per capita output of CO_2 because of its small population (19.7 million) and relatively high CO_2 emissions (410 million metric tons per year). It is not surprising that the United States is the second leading per capita emitter of CO_2 with its enormous output (5.9 billion metric tons per year) but still relatively large population (300 million).

The United States: Superpower and Superpolluter

The United States has the third largest population and the world's largest economy. Because of the large economy its population is mostly affluent and an enormous consumer of goods and energy. It is not surprising that the United States is, by far, the greatest emitter of greenhouse gases ($\sim 25\%$ of the world's emissions). The greenhouse gas emissions by the United States are shown in Figure 8.4. It not only includes CO_2, but also methane, nitrous oxide, and halocarbons. Therefore, the total emission of greenhouse gas equivalent is 1.1 billion metric tons more than CO_2 alone. The total emissions were 6.155 billion metric tons in 1990 and 6.862 billion metric tons in 2002, an increase of 10.3% in the last 13 years. Most of that increase was for CO_2, by far the greatest greenhouse gas produced

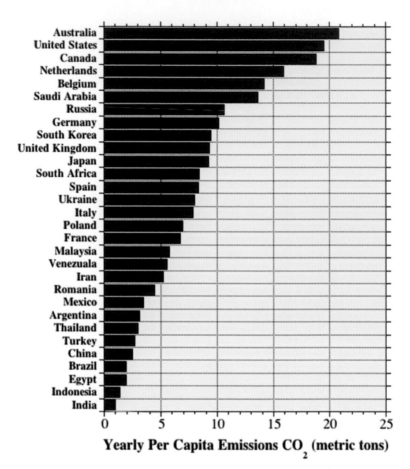

FIGURE 8.3 *This bar charts shows the yearly per capita emissions for various countries. Australia has the highest per capita output because it has a small population but a relatively high emission rate. The second (China) and fifth (India) largest CO_2 emitters have relatively low per capita emission rates because they have such large populations.*

by the United States and the rest of the world. In fact both methane and nitrous oxide declined by 14.8% and 0.2% respectively, but halocarbon emissions increased by 24.6%. The amount of CO_2 absorbed by the land area of the United States in 1990 is estimated to have been 1.073 billion metric tons, but in 2001 that had fallen to an estimated 0.838 billion metric tons, a decrease of 21.8%. Not only has the CO_2 increased by 10.3% in the last 13 years, but the amount absorbed by the land area of the United States has decreased by 21.8%. Therefore, there has been a net increase in the amount of CO_2 equivalent to the atmosphere by the United States of about 32% during that period. The United States is the biggest offender and it is only getting worse.

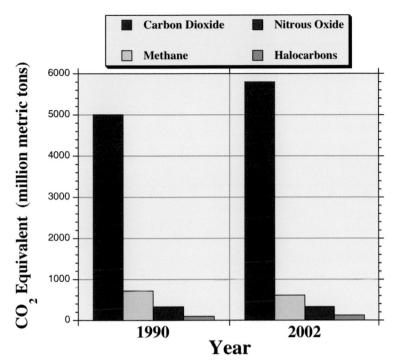

FIGURE 8.4 This shows the United States emissions of the greenhouse gases CO_2, CH_4, N_2O, and halocarbons for 1990 and 2002. See text for discussion. (Source: US Dept of Energy)

Greenhouse gas output can be divided into six sectors: electricity generation, transportation, industrial, agriculture, commercial, and residential. Transportation includes emissions from cars, trucks, and airplanes, while the industrial sector is basically emission resulting from manufacturing. The commercial sector is primarily emissions from retail outlets and other businesses, while the residential sector emissions are primarily due to the consumption of electricity and gas from utilities by residents. Agriculture includes emission from farming the land, including the use of fertilizers.

Figure 8.5 shows the CO_2 emissions for the various economic sectors from 1990 to 2003. By far the largest CO_2 emissions come from generating electricity. In the United States coal is the most common fuel used in power plants. Transportation is the second largest producer of CO_2,. The current (2006) consumption of oil is about 20 million barrels per day, but is estimated to increase to about 27 million barrels per day by 2025. In 2004 greenhouse gas emissions rose to an all time high of 2% over the previous year (US Dept of Energy). This represented 7.1 billion metric tons of CO_2 equivalent, up from 6.98 billion metric tons in 2003.

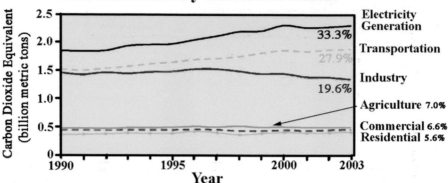

FIGURE 8.5 *This graph shows the United States yearly equivalent CO_2 emissions from 1990 to 2003 for various sectors of the economy. The largest emitter is electricity generation which is heavily dependent on coal (See FIGURE 8.8) followed by transportation (oil). (Source: US Environmental Protection Agency)*

It is projected that the emissions of CO_2 from all sectors will steadily grow during the next 25 years by about 29%, or to 9.16 billion metric tons per year by 2025. The United States is continuing to emit greenhouse gases at a prodigious rate with no signs of concern on the part of the present administration.

China and India: The Awakening Giants

In 1820, China accounted for about 30% of the global economy and India accounted for about 15%. At that time the United States accounted for less than 2% of the world economy. By the 1950s the United States was dominant, and China and India accounted for only about 4% each of the world economy. Today, China and India are two of the fastest growing economies in the world and, between them, account for about 20% of the global economy. The average Gross Domestic Product growth rate is 8% per year in China and 5.8% per year in India. In China, between 250 and 400 million people have been lifted out of poverty by the growing economy and that will surely continue as the economy grows. These newly "affluent" people will also consume more goods and energy, contributing more greenhouse gases to the atmosphere. Together, China and India account for 40% of the world's working-age population. As their economies improve the increased personal income will result in a higher demand for automobiles and therefore greater CO_2 emission. China is also one of the two leading importers of tropical rainforest timber, making it a driving force behind tropical deforestation. It also accounts for 15% of the world fish catch and 33% of the global seafood consumption.

In China, electricity is by far the biggest consumer of energy, but transportation is growing fast. Between 75 and 80% of its electricity is generated by coal, the greatest emitter of greenhouse gases (Liu and Diamond, 2005). Figure 8.6 is a satellite image showing the pollution in northeastern China due primarily to coal burning. Because power demands will increase rapidly, it is probable that coal consumption will also increase dramatically. China lacks large reserves of oil and natural gas, but it has enormous reserves of coal. It is predicted that energy demands will rise to the equivalent of 3.5 billion tons of coal per year by 2020 unless measures to improve efficiency are instituted. China alone has the potential to single-handedly emit enough CO_2 to negate all other nations' efforts to control their greenhouse gas emissions. Although China is a signatory to the Kyoto Protocol it is a developing country that is not yet required to meet an emissions reduction target.

Between 1990 and 2002 greenhouse gas emissions increased 45% in China and almost 71% in India compared to just 16.7% in the United States. In 1993 the annual car sales in India were about 200,000, but in 2004 it more than quadrupled to about 803,000. China's economy alone is expected to quadruple by 2020, and it could be the largest economy in the world by 2030. This means that greenhouse gas emissions from the second largest emitter will increase tremendously. On average, China is building one coal-fired power plant each week. It is now estimated that China will surpass the United States in CO_2 emissions as soon as 2010 (International Energy Agency, 2006). By 2030, China is projected to emit about 11 billion metric tons of CO_2 per year compared to about 8 billion tons per year for the United States (Figure 8.7).

Others

The CO_2 emissions of most other countries have also increased between 1990 and 2002. For example, the International Energy Agency estimates that the increase by France was 6.9%, Italy 8.3%, Greece 28.2%, Ireland 40.3%, the Netherlands 13.2%, Portugal 59%, and Spain 46.9%. However, Germany was down 13.3% and Britain was down 5.5% during this period. Germany decreased its emissions because it closed many inefficient coal-fired plants in eastern Germany after reunification in 1990, and, in Britain, the government shifted electric utilities from coal (high CO_2 emissions) to natural gas (lower CO_2 emissions). These, however, are only one-time reductions.

In Russia only a few industries produce an enormous amount of CO_2.

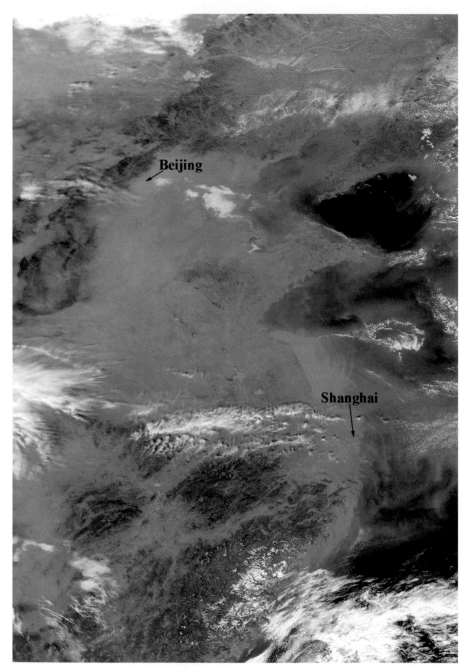

FIGURE 8.6 *This satellite image of China taken November 20, 1999 from the SeaWIFS Project shows the gray pollution primarily from coal burning. Most of the countryside is hidden by the gray smog layer. The position of Beijing and Shanghai are shown by the arrows. (Source: ORB Image, NASA Goddard Space Flight Center, Greenbelt, MD)*

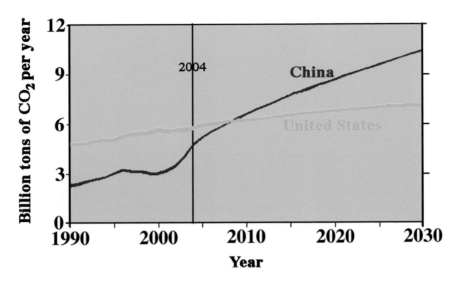

FIGURE 8.7 This graph shows the projected increase in CO_2 emissions from the United States and China beyond 2004. The year 2004 is the last year for which data were available. China is projected to surpass the United States in emission of CO_2 by 2010 unless they drastically reduce their emissions before then. (Modified from International Energy Agency, World Energy Outlook 2006)

For example, Unified Energy Systems, Russia's electricity monopoly, emits 2% of all human-caused emissions of CO_2 or almost the entire output of Britain (World Resources Institute). Under the rules of the Kyoto Protocol they can potentially earn hundreds of millions of dollars by trading their right to release CO_2 into the atmosphere.

ExxonMobil Corporation forecast that global energy demand will rise 50% by 2030 and that oil will remain the dominant fuel source. According to them electricity generation is the fastest-growing energy need, and that it will boost the demand for coal and natural gas. This is consistent with the IEA World Energy Outlook 2004 prediction of a 63% increase in CO_2 emissions over the 2002 levels by 2030. Therefore, without urgent and very strong decreases in CO_2 over the next 20 years we can expect a temperature rise of between about 0.5 and 2 °C over today's level by 2050 (see Chapter 12 for a more thorough discussion).

COAL: THE WORST FOSSIL FUEL OFFENDER

There are three types of fossil fuel; oil, natural gas, and coal. By far the worst fossil fuel for the emission of greenhouse gases is coal. Coal produces more CO_2 per energy unit than any other fossil fuel, and there are about a trillion tons available for mining. Anthracite coal contains about 92% carbon and brown coal about 70%. Burning anthracite coal results in about 4 tons of CO_2 for each ton of coal consumed, because the emitted carbon combines with atmospheric O_2 to form CO_2; and coal-fired power plants consume huge amounts of coal, some over 500 tons an hour. In these cases over 2,000 tons of CO_2 are released each hour from these power plants.

Most of the readily available coal reserves are in the United States, Russia and China (Figure 8.8), and the annual world consumption is about 5 billion tons, or the average emission of over 10 billion tons of CO_2 each year. Figure 8.9 shows the present leading coal burners, the amount they consume and their projected consumption for 2025. The United States emits more CO_2 by generating electricity than any other economic sector (see Figure 8.5). Figure 8.10 shows the United States percent energy source for electricity generation and their power plant percentage CO_2 emissions. Today United States power plants release all of their CO_2 into the atmosphere. Nearly 2 billion tons of it are deposited in the atmosphere each year from United States coal-fired power plants, and that figure could rise by about 30% in 20 years. It is obvious from these graphs that the world has a real problem with CO_2 emissions unless coal

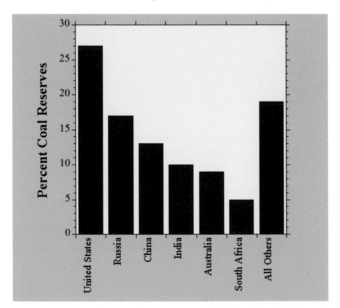

FIGURE 8.8 The percentage of world coal reserves available for mining by the six leading countries is shown on this graph. The United States, Russia and China have the most readily available coal. (Source: Energy Information Administration, US Dept of Energy)

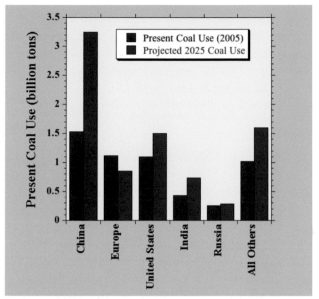

FIGURE 8.9 This graph shows the present consumption of coal and projected use in 2025. China is projected to have the greatest increase, but with substantial increases for the United States, India and all other countries combined. Europe is expected to have a decrease and Russia only a small increased usage. (Source: Energy Information Administration, US Dept of Energy)

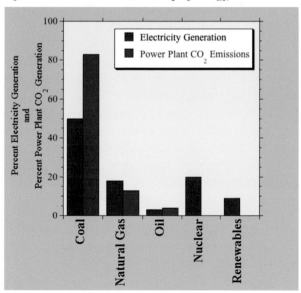

FIGURE 8.10 This graph shows the percent electricity generation from coal, natural gas, oil, nuclear energy, and renewable resources for the United States. It also shows the percent of CO_2 power plant emissions. Coal is the greatest producer of electricity and also the greatest producer of CO_2 of any power source. Nuclear energy and renewables do not produce CO_2. (Source: Energy Information Administration, US Dept of Energy)

emissions are brought under control. It is perfectly feasible to built a coal-fired power plant that does not release greenhouse gases to the atmosphere, but to date we are doing little to implement such conversions, and some of the ways of sequestering CO_2 from power plants are discussed in Chapter 13.

SUMMARY

We are pouring greenhouse gases into the atmosphere in prodigious quantities and if we do nothing to curb these emissions we are going to be in serious trouble. In particular, the leading emitters of greenhouse gases shown in Figure 8.2 need to take the lead in reducing these gases. These 10 countries combined, account for about 70% of the world's emissions of greenhouse gases, about the percentage reduction needed to stabilize the atmospheric greenhouse gas content. The United States, in particular, needs to set an example by vigorously pursuing a policy for reducing greenhouse gas emissions. Unfortunately, the major polluter, the United States, has refused to ratify the Kyoto Protocol to limit the emission of greenhouse gases. Although China, Russia, and India (the second, third, and fifth largest emitters) signed the Protocol, they do not have to limit emissions: in Russia, current emissions are less than they were in 1990, and as China and India are developing countries they are not bound to limit their emissions in this round of the Protocol. Thus, four of the five largest emitters will be doing nothing to limit their greenhouse gas emissions under the Kyoto Protocol because they either have not ratified it or are not required to reduce their emissions under the current rules. It would still not be enough to stabilize atmospheric CO_2 if the 10 largest polluters continued emitting greenhouse gases at their current rate, and all other countries of the world cut their emissions to zero.

9

THE ROOT CAUSE

EACH one of us is responsible for the release of greenhouse gases into the atmosphere. Every time we drive a car, turn on a stove or a light, heat or cool a house, or contribute waste to a landfill, we are directly or indirectly emitting greenhouse gases (mainly CO_2). Even just breathing emits 1 kilogram of CO_2 each day by every person on Earth. Our lifestyle determines how much greenhouse gases we emit. An affluent society like the United States emits enormous quantities of greenhouse gases, while non-industrialized third world countries emit much less. On average, each person in the United States emits almost 20 metric tons of CO_2 each year, but in Bangladesh, for example, each person emits only about 0.02 metric ton each year. An average American, therefore, emits 1,000 times more CO_2 than an average Bangladeshi. If you are a rich American who lives in a large house using lots of energy and you drive a truck-like, gas-guzzling Sports Utility Vehicle (SUV) that only gets about 10 miles per gallon, then you are probably emitting over 40 metric tons of CO_2 each year. In other words, the amount of CO_2 people emit depends on where they live (industrialized or non-industrialized country) and their lifestyle (rich

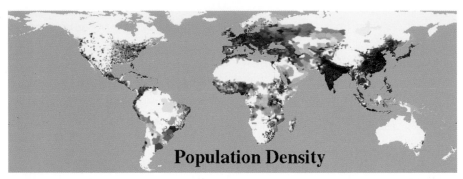

FIGURE 9.1 *This map shows the world's population density. (The redder the region the higher the population density.) In general the most populated regions are Europe, India, China, Japan, and Java Indonesia. (Data from the UN Population Reference Bureau)*

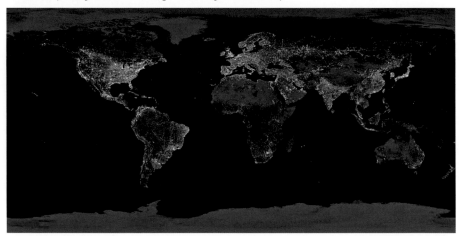

FIGURE 9.2 *This map shows the distribution of lights on the Earth. The concentration of lights generally coincides with regions of large population or high industry. Notice that the highly industrial countries of the United States, Europe, Japan, and South Korea have very high concentrations of lights. (From US National Oceanic and Atmospheric Administration, Earth Observation Group, National Geophysical Data Center)*

or poor). It is obvious that the more people there are, the more greenhouse gases are emitted to the atmosphere.

Today (2007) there are over 6.6 billion people on Earth and another 4.4 billion domestic animals (UN Population Division). Humans and their domestic animals account for between 85 and 90% by weight of all vertebrates on Earth (MacKenzie, 1998)—that includes mammals, birds, fishes, and all other creatures with vertebrae. Although an elephant, whale, or other large vertebrate weighs much more than a human, there are very few of them compared to humans. The 6.6 billion people are not distributed evenly across the continents, as can be seen in Figure 9.1 which shows the current population density (number of people per unit area) on Earth. The greatest population densities are in parts of Europe, India, China, Japan, and the island of Java. The amount of industrialization and population density are generally shown graphically by the distribution of lights at night (Figure 9.2), where the greater concentration of lights usually represents the more industrialized regions. You can see in Figure 9.2 that the industrial giants, United States, Europe, Japan, and South Korea, for example, are crammed full of lights. They are large greenhouse gas emitters. The populous, but unindustrialized, North Korea has almost no lights.

HUMAN POPULATION GROWTH

The uncontrolled growth of the human population and industry during the past 200 years is the root cause of today's global warming and many other environmental problems. Figure 9.3 shows the growth of the human population from the very beginning of civilization 12,000 years ago to the present (UN Population Division). It is estimated that 12,000 years ago there were only about 1 to 5 million people on Earth. From 10,000 BC to

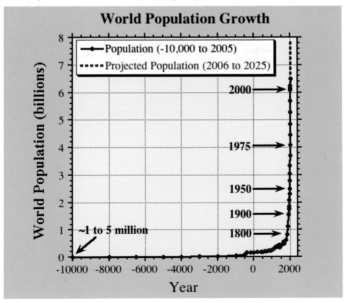

FIGURE 9.3 This graph shows the world population growth from 12,000 years ago through 2005. The dashed red line is the projected population growth between 2006 and 2025. The growth has been exponential since about 1800. (Data from the UN Population Reference Bureau)

about 1750 AD, a period of 11,750 years, there was only gradual growth because of the high death rate and short life expectancy. It took approximately 11,800 years before the population reached 1 billion people in about 1800. At that time the population was beginning to grow exponentially because of improved health care, growing birth rates and a longer life expectancy. In the next 100 years (1800–1900 AD) the population grew to 1.5 billion, and in the following 100 years (1900–2000 AD) the population soared to 6 billion. Therefore, it took 11,800 years for the population to reach 1 billion, but in modern times only 200 years to reach 6 billion.

There is a slight dip in population at about 1300 AD that coincides with the Black Death scourge in Europe, and a slowing down of population growth at the time of the First World War and its 10 million deaths, but nothing shows up in the 1940s when there were 35 million deaths from the Second World War. The population growth rate at that time was just too high for the war to make a dent in the population curve. In my 73-year lifetime I have seen the world population increase by 4 billion people, which is twice as much as during the previous 11,933 years. The five countries with the largest yearly population increases between 2000 and 2005 are India (16 million), China (9.4 million), Pakistan (3.7 million), Nigeria (3.1 million), and the United States (3.0 million).

Today's population of 6.6 billion people is pouring out greenhouse gases at the prodigious rate of about 25 billion metric tons per year, or an average rate of almost 4 metric tons of CO_2 per person per year. There are direct correlations between the growth of the world population, the increase in the atmospheric CO_2 content (Figure 9.4), and the human-caused emission of CO_2 (Figure 9.5). The graphs in these figures show that the rate of increase of CO_2 and its emission by humans is almost exactly the same as the rate of population growth. There is no doubt that the growing human population together with growing industrialization is the primary cause of the rising CO_2 emissions, the rise in the atmospheric CO_2 abundance, and the increase in the global temperature.

Another consequence of the large human population and their domestic animals is that they emit a large amount of CO_2 by just staying alive. Each person exhales about 1 kilogram of CO_2 each day, and larger domestic animals presumably exhale at least that much if not more. The human population alone exhales about 2.3 billion metric tons of CO_2 each year, and just the large domestic animal population of about 1.6 billion exhales another 5.8 million metric tons of CO_2 each year. The combined amount of CO_2 exhaled from humans and their larger domestic animals is about 2.9 billion metric tons each year. Normally this amount

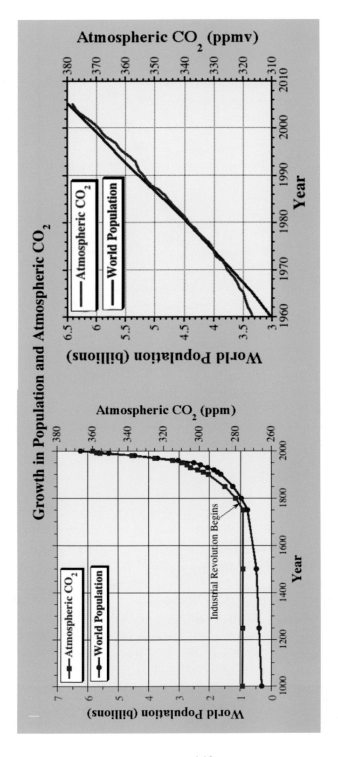

FIGURE 9.4 These graphs show the population growth and increase in atmospheric CO_2 from 1000 AD to 2002 on the left, and the same for the period 1960–2006 on the right. There is an excellent correlation between the population and CO_2 curves indicating that the rate of increase in atmospheric CO_2 and the rate of population growth are the same. This is further evidence that humans are the cause of the rise in atmospheric CO_2.

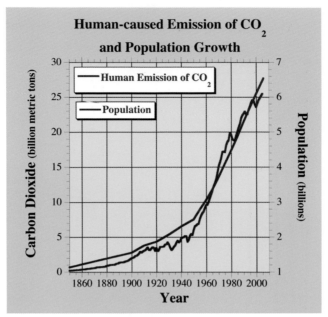

FIGURE 9.5 This graph shows the population growth and the increase in CO_2 emissions from 1850 to 2005. The good correlation indicates that the rate of increase of human-caused CO_2 emissions and population growth are the same.

of CO_2 would be easily absorbed by the Earth. However, the CO_2 exhaled by humans and their domestic animals becomes significant when the additional large amount of CO_2 resulting from burning fossil fuel and land use is taken into account.

THE NEXT 25 YEARS

Although the birth rate is currently falling, the world population is expected to grow to over 9 billion during the next 50 years (UN Population Division). The projected large increase in life expectancy more than offsets the declining birth rate. Predicting the population 50 years from now is uncertain because many things could change in that time, and the predicted population growth over the next 25 years (dashed line in Figure 9.3) is probably more reliable. By 2025 the population is predicted to grow by 1.852 billion to a total of 8.05 billion people (UN Population Division), and the expected population increase or decrease for various parts of the world can be seen in Figure 9.6. Because of the declining birth rates in Europe, Russia, and Japan, their populations are predicted to fall a

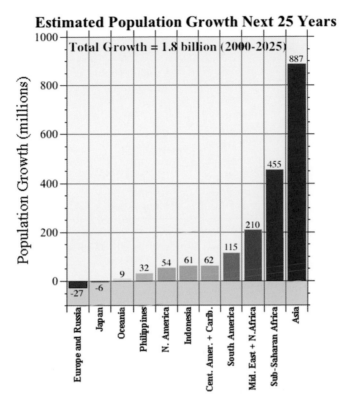

FIGURE 9.6 *This bar chart show the projected population growth from 2000 to 2025 for various regions of the world. Europe, Russia and Japan lose population but the rest of the world gains. The biggest increases are in Asia (excluding Japan), Sub-Saharan Africa, and the Middle East and North Africa. (Data from the UN Population Reference Bureau)*

total of 33 million in the next 25 years. The largest increases occur in Asia (excluding Japan), sub-Saharan Africa, the Middle East, and North Africa. In these regions the total increase is about 1.55 billion or about 84% of the total 25-year growth. Some people believe that the AIDS epidemic in southern Africa will result in a decreasing population, but that is not true because of the very high birth rate there (Table 9.1).

The predicted population growth in the next 25 years will add appreciably to the CO_2 content of the atmosphere. Assuming an average per capita emission rate the same as today (4 billion metric tons per person per year), by 2025 we will be adding each year about 7.4 billion metric tons more CO_2 than today unless something is done to curb greenhouse gas emissions. In 2025, all other things being equal, we will still be emitting the same amount of CO_2 even if we reduce our current emission rate by 25%.

TABLE 9.1 *Ten countries or areas with the highest and lowest fertility rates, 2000–2005*

Highest total fertility[a]		Lowest total fertility[a]	
Niger	8.00	Czech Republic	1.16
Somalia	7.25	Armenia	1.15
Angola	7.20	Spain	1.15
Guinea-Bissau	7.10	Ukraine	1.15
Uganda	7.10	Russia	1.14
Yemen	7.01	Slovenia	1.14
Mali	7.00	Bulgaria	1.10
Afghanistan	6.80	Macao, China	1.10
Burundi	6.80	Latvia	1.10
Liberia	6.80	Hong Kong, China	1.00

[a] Average number of children per woman.
Note: The world's average total fertility is 2.69. A fertility rate of 2.1 is required to stabilize the world population (zero population growth).
Source: United Nations Population Division.

OTHER ENVIRONMENTAL PROBLEMS

Although mostly unrelated to global warming, the current large and growing population is having other serious affects on the environment. Combined with the consequences of global warming, the environment will be further stressed enormously. The United Nations sponsored Millennium Ecosystem Assessment (2006) has studied the impact of the human population on the environment. This assessment group consisted of 1,360 scientists from 95 countries working over a four-year period. They found that the growing population and expanding economic activity during the past 50 years have depleted 60% of the Earth's grasslands, forests, farmlands, rivers, and lakes. They also found that in recent decades one-fifth of the coral reefs and a third of the mangrove forests have been destroyed, the diversity of animal and plant species have declined sharply, and a third of all species are at risk of extinction. Disease outbreaks, floods, and fires have become more frequent. Some of these events are, indeed, due to global warming; humans are now changing ecosystems more rapidly and extensively than at any time in human history.

Another example is the increasing fish catch. To feed a growing population the fish catch has grown in direct proportion to the population (Figure 9.7). This has led to an enormous decrease in the food fish population. Fifteen years after the initiation of industrialized fisheries the

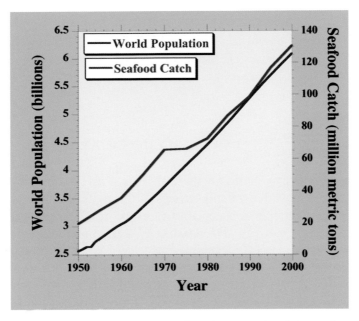

FIGURE 9.7 *This graph shows the correlation of population growth and the growth in the fish catch between 1950 and 2000. The decreasing fish population requires fishing boats to go farther out to sea for fish using more oil and emitting more CO_2. (Fish data from Myers and Worm, 2003)*

commercial fish population was reduced by 80%. It is estimated that the global ocean has lost more than 90% of large predatory fish (Myers and Worm, 2003). For example, the cod population in the North Atlantic Ocean has decreased 96% during the past 50 years due to over-fishing. A new study (Worm et al., 2006) indicates that if commercial fishing and ocean pollution continue at the current rate, by 2050 there will be a "global collapse" of all species currently fished. "Collapse" is defined as less than 10% of a species surviving. This could be averted if we take action now to curb global over-fishing as has been done by the United States in its coastal waters.

You may think that over-fishing is unrelated to global warming, but you would be wrong. Because of the dwindling fish supply, fishing boats have to venture farther and farther out to sea to find their catch. In doing so they are using much more fuel than before and emitting much more CO_2. In 2000, fishing fleets harvested about 80 million tons of fish and burned about 50 billion liters of fuel (Tyedmers et al., 2005). That amount of fuel was about 1.2% of global oil consumption.

There are many other environmental problems due to the large human population that are beyond the scope of this book. All of these environmental woes are happening at the same time that global warming is producing its own environmental problems. Anyone who thinks the world population is not seriously affecting our ability to survive the future is living in a "fool's paradise."

THE EARTH'S CARRYING CAPACITY

How many people can the Earth sustain? This is a very difficult question to answer. There have been many studies of this problem, but no one has suggested an answer that is acceptable to everyone. The problem is that technology keeps changing at a rapid rate, and, therefore, it is hard to predict new technological developments that might help to maintain human sustainability. In the 1960s it was predicted that the world's ability to feed itself would come to an end in the 1980s because the world's population would outgrow the available food supply. They did not consider the changing technology that led to the "Green Revolution" and the many-fold increase in food production. Today genetic engineering of crops may lead to a new "Green Revolution" that could provide more food to a growing population. Possibly fish farming, at least of some species, could make up for some of the depleted fish supplies mentioned above. However, the disruption of the environment, including global warming, by the current large population could indeed catastrophically affect human survival. People who study the Earth's carrying capacity have mixed opinions on our sustainability. The optimists say we are very close to the Earth's carrying capacity, and the pessimists say we have already exceeded it by a factor of 2. In other words, in this more pessimistic view we need two Earths to sustain the current human population. All this may be moot if global warming continues unabated.

SUMMARY

Each of us is responsible for the release of greenhouse gases to the atmosphere because of our energy consumption and other factors. The root cause of the present warming is the Industrial Revolution and the accompanying rapid rise in the world population. There is a strong relationship between the growth of the population and the growth of human-caused greenhouse gas emissions. This same relationship also exists for the increase in greenhouse gas content of the atmosphere. It is estimated that the next 25 years will see a 1.8 billion increase in the world population to almost 8 billion people. Each of these people will be causing a certain amount of greenhouse gas emission that will make it more difficult to curb future emissions. This makes it even more urgent to begin now a program of curbing greenhouse gas emissions.

10

THE MELTING
EARTH

MUCH of the Earth's ice is melting at an unprecedented rate. If we had no other information than this, we would be forced to conclude that the Earth was warming. The Earth's ice deposits can be divided into four broad groups: (1) mountain glaciers, (2) sea ice, (3) ice sheets, and (4) ground ice in permafrost. All four types are left over from the end of the last ice age. During the Hot House periods in Earth's history there were no accumulations of ice.

Mountain glaciers occur in certain mountainous terrain at all latitudes, including high mountains in the tropics. Ice sheets cover broad areas to considerable depth. There are only two ice sheets: one covers the Antarctic continent at the South Pole, and the other covers most of Greenland near the Arctic Circle. Sea ice, which is the frozen surface of oceans or seas, forms ice shelves. The area of sea ice depends on the seasons, and are more extensive in winter than they are in summer. The main ice shelves occur in both the Arctic and Antarctic regions.

Ground ice is ice occurring in permafrost or permanently frozen soil. Permafrost is found at high latitudes mainly in the Arctic regions because that is where most of the high-latitude land surface occurs. It underlies about 24% of the land surface in the Northern Hemisphere, or the equivalent of approximately 22.79 million km^2 (Zhang, 2003). Permafrost can also contain up to 50% ground ice. Both ground ice and permafrost are sinks for CO_2 and methane and together they store about 14% of the world's carbon. Permafrost, which is no longer permanent and is now beginning to melt, occurs on land and below the ocean on Arctic continental shelves. Its thickness ranges from less than 1 meter to greater than 1,000 meters.

Ice deposits act in a complicated manner that depends on a variety of factors. They can advance or retreat due to the environment in which they occur, particularly in estuaries. As far as the climate is concerned, the main factor is the mass balance of glaciers, ice sheets, and sea ice. If the mass of these ice deposits is increasing then there is probably more precipitation and accumulation of ice at their heads than is melting at their ends. On the other hand, if the mass is decreasing, as most of them are today, then they are melting faster than ice is accumulating at their heads. Temperature, precipitation, humidity, wind speed, and other factors such as slope and the reflectivity of the glacier surface all affect the mass balance of a glacier. However, most glaciers are more sensitive to temperature than other climatic factors (Fitzharris et al., 1996).

Mountain glaciers and small ice caps cover 680,000 km^2, and the water contained in this ice would raise sea level about 0.5 meter. The area covered by the Greenland and Antarctic ice sheets combined is 14.1 million km^2 (1.71 million for Greenland and 12.37 million for Antarctica), and contains enough water to raise the sea level by 70 meters (8.6 meters for Greenland and 61.4 meters for Antarctica). This sea level rise does not include any rise due to thermal expansion of the oceans.

MOUNTAIN GLACIERS

Most mountain glaciers are melting at a high rate. For 30 glaciers in nine regions of the world, there has been a cumulative net loss of 9 meters thickness since 1985 (World Glacier Monitoring Service). This is equivalent to more than 6,700 cubic kilometers of water. Figure 10.1 graphically shows the loss up to 2004. Recent analyses of satellite data also indicate that global glacier decline has been accelerating since the 1980s.

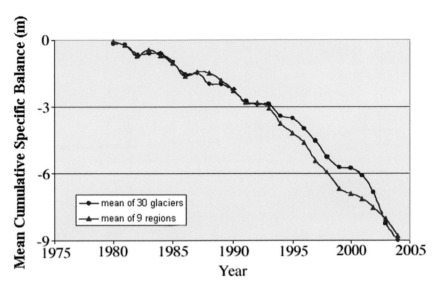

FIGURE 10.1 *This graph shows the decrease in the mass of glaciers from 1980 to 2004. The average decline in glacial thickness has been 9 meters in this 24-year period. (Courtesy of World Glacier Monitoring Service)*

The graph in Figure 10.1 shows a particularly rapid acceleration since 2001 for the 30 glaciers. Figure 10.2 shows examples of glacial melting in the United States, Canada and Peru, and glaciers are continuing to melt at an accelerated rate. A recent study (Kaser et al., 2006) indicates that the world's glaciers and small ice caps are now in "terminal decline," and that "99.9% of all glaciers" are now shrinking because of global warming.

Alpine glaciers are probably the best-studied glaciers in the world. They have been melting and retreating since 1850. Recent analyses show they are rapidly melting and that the melting is accelerating—For example, the famous "Mer de Glace" glacier on Mont Blanc had a thinning rate increase of 1 meter per year between 1979 and 1994, 2.9 meters per year between 1994 and 2000, and 4 meters per year between 2000 and 2003 (Berthier et al., 2004). A study of about 930 Alpine glaciers shows an 18% reduction in glacial area between 1985 and 1999, which is a seven times higher loss rate than for the period between 1850 and 1973 (Paul et al., 2004). This corresponds to a volume loss of 25 km^3 since 1973 and there will be a positive feedback because of increased thermal radiation from now ice-free areas that will further accelerate glacial melting. Furthermore, the response time for Alpine glaciers to climate changes is about 10–20 years, so the accelerated melting due to the extraordinarily warm 1990–2000 decade is still to be experienced. At the current rate of loss, all Alpine glaciers will be gone in about 75 years, and with a continued acceleration

FIGURE 10.2 These images show examples of glacial melting in the USA, Peru, and Canada. This decline is continuing at a rapid rate. (Sources: Glacier National Park (Alden/GNP Archives [1913]; Blase Reardon/US Geological Survey [2005]. Qori Kalis Glacier (Lonnie Thompson, Ohio State Univ.). Athabasca Glacier (BC Provincial Boundary Commission [1917]; B.H. Luckman [1986])

of the loss rate they will be gone much sooner. Many Alpine ski resorts will be in financial trouble as their ski areas are reduced. In May of 2005, more than 3,000 square meters of the Gurschen glacier above the Swiss resort of Andermatt were covered with an insulating foam to protect the glacier from melting. Other Swiss resorts, including Saas-Fee and Engelberg-Titlis, and ski centers in Austria and Italy, are considering similar measures.

- In Peru the Qori Kalis glacier was retreating at a rate of 4.7 meters per year in 1978, but in 2001 its retreat rate had increased to 205 meters per year; an average rate increase of 8.7 meters per year (Figure 10.2).

- In 1850 there were 150 glaciers in Glacier National Park, Montana. Today only 50 remain so that the area covered by glaciers has declined by 73% from 1850 to 1993. The largest glaciers are about one-third of their size in 1850. Today the total area covered by all glaciers is 27 km^2 compared to the 99 km^2 that previously existed. At the current rate of melting, there will be no glaciers within 30 years (Figure 10.2).

- A study of 67 glaciers in Alaska from the mid-1950s to the mid-1990s indicates an average decline in thickness of 0.52 meter per year (Arendt et al., 2002). Extrapolation to all Alaskan glaciers suggests an annual volume decrease of 52 km^3 per year. Repeat measurements of 28 glaciers for the mid-1990s to 2001 indicated that the thinning rate had increased to 1.8 meters per year. This suggests that the loss of glacier volume in Alaska by 2001 had increased to 96 km^3 per year. On the Malaspina Glacier, there was a 20–25 meter decrease in elevation between 2000 and 2004 (Sauber et al., 2005), and the Harding Icefield and Grewingk-Yalik Glacier Complex are thinning at a rate of about 0.61 m/yr (VanLooy et al., 2006). The melting of Alaska's glaciers alone would result in a sea level rise of 0.27 mm/year during the past decade. (Figure 10.3 shows the change of three Alaskan glaciers due to melting. It also shows the change in vegetation from a colder to a warmer climate.)

- Glaciers in New Zealand's Southern Alps have lost 25% of their area over the past 100 years (Chinn, 1996).

- In the Himalayas glaciers are melting at comparable and accelerating rates as other world glaciers (Kargel et al., 2005). The area covered by glaciers in the Ak-shirak range of Central Asia shrank 3.4% between 1943 and 1977, but more than 20% between 1977 and 2001 (Khromova et al., 2003). The largest number and coverage of mountain glaciers occurs on the Tibetan Plateau, where there are over 46,000 glaciers covering almost 155,000 km^2. The average altitude of the glaciers is about 3,900 meters, and according to the Chinese Academy of Sciences, the glaciers are decreasing at a rate of 7% each year—enough meltwater to fill the entire Yellow River. They believe the environmental changes caused by this melting will further increase droughts and sandstorms over the rest of China (see Chapter 11). Although the glaciers have been receding for the past 40 years, the process is accelerating at an alarming rate due to escalating temperatures. The temperature in Tibet has increased over 1 °C during the past 20 years.

Aug. 1941   **Aug. 2004**

Aug. 1906   **Sept. 2003**

July 1909   **Aug. 2004**

FIGURE 10.3 These images show the decline of three Alaskan glaciers due to melting. The top images are of the Muir Glacier, the middle images are of the Carrol Glacier, and the bottom images are of the McCarty Glacier. In addition to the ice decline they also show the dramatic change in vegetation from a colder to a warmer climate in the top two comparisons. (The images were taken by W.O. Field (upper left), C.W. Wright (middle left), US Grant (lower left), and B.F. Molnia (all right images). NSIDC/WDC for Glaciology, Boulder, compiler, 2002, updated 2006. Online glacier photograph database. Boulder, CO: National Snow and Ice Data Center/World Data Center for Glaciology. Digital media)

- Probably the most famous glaciers in the world are on Mt Kilimanjaro near the equator in Africa. Kilimanjaro has lost 82% of its glaciers since 1912 (Figure 10.4), and in about 15 years the Kilimanjaro glaciers will be gone. Global warming plus other environmental changes such as forest clearing that reduces rainfall are leading to the imminent extinction of probably the most famous glaciers on Earth. Also the

FIGURE 10.4 (Left) This NASA space shuttle image shows the small area covered by glaciers on Mt Kilimanjaro in 2000. (Right) The graph shows the decrease in the area of the Kilimanjaro glaciers between 1912 and 2001. At the current rate of decline they will be gone in a few years. (Graph from Alverson et al., 2001)

157

glaciers in the Rwenzori Mountains of East Africa are losing mass at a rapid rate due to increased air temperatures (0.5 °C per decade) over the past 40 years (Taylor et al., 2006). The glaciers lost about half their area between 1987 (2 km^2) and 2003 (0.9 km^2), and are expected to disappear completely within the next 20 years.

These are just a few of the examples of mountain glacial melting. Most of the world's mountain glaciers are melting and retreating. In some mountain communities, particularly in the Andes, this melting will have significant socioeconomic impacts from alteration of the hydrology and eventually from waning water supplies from the shrinking glaciers (Bradley et al., 2006). In a warmer world, less winter precipitation is snow, and snow melting occurs earlier in spring. This leads to a shift in peak river runoff to winter and early spring and away from summer and autumn when demand is highest. Therefore, without adequate storage capacities, most of the winter runoff is lost to the oceans. Areas dependent on this runoff are likely to experience severe water shortages (Barnett et al., 2005).

Areas dependent on tourism will suffer from uncertain snow cover during peak winter sports seasons, increased natural hazards such as rock and ice-falls, or just plain loss of beauty. During the 2003 summer heat wave in Europe people had to be airlifted off Alpine trails because of the danger of falling rocks dislodged from rapidly melting glaciers.

THE ARCTIC AND GREENLAND

The Arctic

Thirty years ago in Fairbanks, Alaska, temperatures reached 26 °C (80 °F) for only one week, but today it reaches or exceeds that temperature for about three weeks. The number of −40 °C winter temperatures has also decreased in the last 30 years. Sixteen years ago cruise ships used to come close to the Columbia glacier's 60-meter-high front in Prince William Sound to watch the ice break off and crash into the sea. Today that is impossible because the front has retreated more than 13 km, forming a great expanse of melting icebergs. Large areas of forest are drowning, turning gray, and trees and roadside utility poles are tilting at odd angles because of melting permafrost. Mountainsides of spruce forest are dead because they were weakened by climate-related stresses and then killed by bark beetles whose population has exploded because of higher temperatures. These are only a few of the consequences of global warming in the Alaskan Arctic.

Arctic regions are warming faster than any other region in the world (Polyakov et al., 2005). The Arctic Climate Impact Assessment study (2004)—a four-year comprehensive scientific assessment (over 1,800 pages long)—found that there have been serious and accelerating impacts on the Arctic environment by global warming. Over the past 50 years, average Arctic temperatures have increased twice as much as the global average and some parts of the Arctic region have experienced considerably greater increases (Poulsen, 2004). In Alaska and western Canada, average winter temperatures increased by as much as 3 to 4 °C over the past 60 years. This is a highly significant increase because the average increase in the Northern Hemisphere over the past 100 years has been only about 0.6 ± 0.2 °C. Furthermore, computer projections indicate that the Arctic will continue to experience more and accelerated warming than the rest of the world for at least the next 100 years (Figure 10.5). The result of these increased temperatures has caused increases in surface and oceanic temperatures, an overall increase in precipitation that is more evident in some sub-regions of the Arctic than in others, large reductions in sea ice and glacier volume, increases in river runoff and sea level, the thawing of permafrost, and shifts in the ranges of plant and animal species.

The Arctic sea ice is rapidly decreasing in thickness and extent (Comiso, 2002; Stroeve et al., 2005). The average summer (July through September) Arctic sea ice extent has decreased for the past 30 years. However, the decrease has been 20% faster over the past two decades (Lindsay and Zhang, 2005). The summer of 2005 saw the most dramatic decline of sea ice (18.2% loss). The loss exceeded the 1979–2000 average by an area of 5.35 million km^2 or about the size of Texas, and was 670,000 km^2 below the previous record low in 2002. The rate of September Arctic sea ice decline from 1979 through 2001 was about 6.5% per decade, but in 2005 the decline had increased to 8% per decade. Figure 10.6, reconstructed from satellite data, shows the decline in sea ice from 1979 to 2003.

There is a major atmospheric circulation pattern called the Arctic Oscillation that can have either a positive or a negative trend. The positive mode creates winds that partially break up the sea ice and flush it out of the Arctic Ocean. Thinner ice is left behind and is easier to melt. In the early to mid-1990s the Arctic Oscillation was strongly positive, which may have made the sea ice more vulnerable to melting. However, since the late 1990s the Arctic Oscillation has been neutral, so it is likely that the decreasing Arctic sea ice is due to global warming. Also, the recent decline in the extent of winter sea ice reinforces that conclusion.

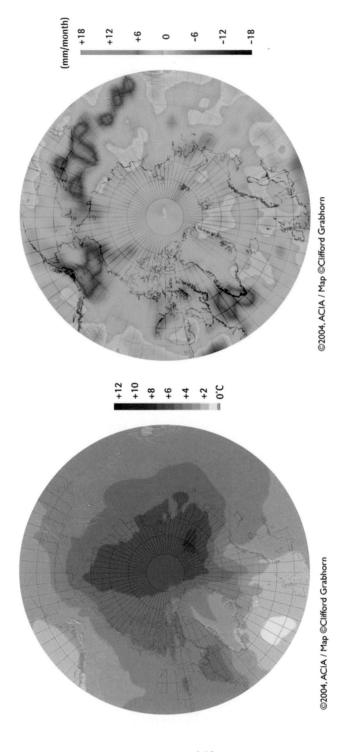

(mm/month)

+18
+12
+6
0
-6
-12
-18

©2004, ACIA / Map ©Clifford Grabhorn

+12
+10
+8
+6
+4
+2
0°C

©2004, ACIA / Map ©Clifford Grabhorn

FIGURE 10.5 These maps show the projected annual surface air temperature change in the Arctic regions from the 1990s to the 2090s (left), and the projected August precipitation change from 1980-1999 to 2070-2089 (right). (From the Arctic Climate Impact Assessment Report, 2004)

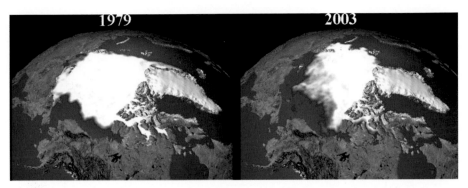

FIGURE 10.6 *Arctic summer sea ice is decreasing at a rate of 8% per decade. The left image shows the minimum sea ice concentration for 1979, and the right image shows the minimum sea ice concentration in 2003. The data used to create these images came from the Defense Meteorological Satellite Program (DMSP). (Source: Krishna Ramanujan, NASA Goddard Space Flight Center, Greenbelt, Md.)*

Although sea ice melts in the summer and reforms in winter, it has also recently failed to reach its maximum extent in the winter (Meier et al., 2005; Comiso, 2006). Since about 1980 there has been a steady decline of about 3% in the area of sea ice during the winter season (December through April); however, all months in the winter and spring of 2004–2005 were well below normal (Figure 10.7). Through March 2006 the loss of winter ice was 300,000 km^2 more than the previous year, as shown in Figure 10.7. Another unusual circumstance is that the reduction occurred in all regions of the Arctic on both the Atlantic and Pacific sides. In the past, one area of the Artic would be unusually low while another area would be unusually high.

Computer model simulations project substantial future reductions in summertime sea ice around the entire Arctic Basin, and one model even predicts ice-free Arctic summers by the middle of this century; however, these models may, in fact, be optimistic. The summer of 2005 was the fourth in succession in which the sea ice has fallen below the monthly downward trend.

The current trend of rapidly declining sea ice could indicate that we have crossed a critical threshold beyond which the climate may never recover. It is now possible that the Arctic has entered an irreversible phase of warming that will rapidly accelerate the loss of the Arctic sea ice (see, for example, Overpeck et al., 2005). As the ice melts it is exposing more dark surfaces (ocean) that absorb more heat causing the ice to melt still further, reinforcing a cycle of melting and heating (positive feedback). Another positive feedback is also occurring due to changes in the reflectivity of the ground. A recent study of summer ground surface darkening in Alaska shows that it is contributing substantially to high-latitude warming trends

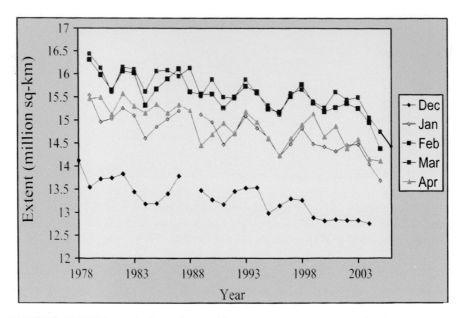

FIGURE 10.7 *This graph shows the monthly mean Arctic sea ice extent for the winter season (December to April) since 1978. All months set record lows for the 2004–2005 winter season. There has been a particularly sharp decline over the past three years in January through March, and the extension of the March curve to 2006 shows the 300,000 km² lost in that winter. (Modified from Meier et al., 2005)*

(Chapin et al., 2005). The pronounced ground surface warming is due to the lengthening of the snow-free season that has caused a warming of the local atmosphere by about 3 W/m² each decade. This positive feedback is expected to continue and could increase atmospheric heating two-fold to seven-fold.

The Arctic heating trend is having profound effects on the ecosystem, and it will have major effects on the weather in the Northern Hemisphere. For example, polar bears depend on the presence of sea ice where they hunt seals and use the ice as corridors for moving between areas. Their success in hunting seals is essential for their survival and depends on good spring ice conditions. As the spring ice shrinks, hunting becomes much more difficult and threatens the existence of polar bears. Sometimes they break through the thin ice, and the rate of starvation is increasing because of their inability to catch seals on the shrinking ice. Other adverse impacts are occurring for Caribou and reindeer, seal breeding, and native human populations (Arctic Climate Impact Assessment, 2005).

Greenland

The Greenland ice sheet is the largest land ice mass in the Northern Hemisphere (1.7 million km^2 or about 9% of all ice on Earth). It averages 1.7 km thick and has a volume that, if melted, would raise sea level 8.6 meters. Although the margins of Greenland are melting, there has been an accumulation of snow in the interior of both Greenland and Antarctica. Between 1992 and 2003 in Greenland there has been a 6.4 cm/year increase above an elevation of 1.5 km, but a 2.0 cm/year decrease at elevations below 1.5 km (Johannessen et al., 2005). Ironically this interior increase is probably due to global warming because warmer air holds more moisture from the increased evaporation of the warming oceans. This is probably causing increased precipitation in the interior of both Greenland and Antarctica.

Satellite data show that between 1992 and 2005 the melt regions of Greenland expanded about 18% (Figure 10.8). Figure 10.9 shows the expansion of the melt zone on one part of the ice sheet between 2001 and 2003. The average ice loss between 1992 and 1997 was 60 km^3/year, but from 1997 to 2003 the loss had increased to 80 km^3/year (Krabill et al., 2004).

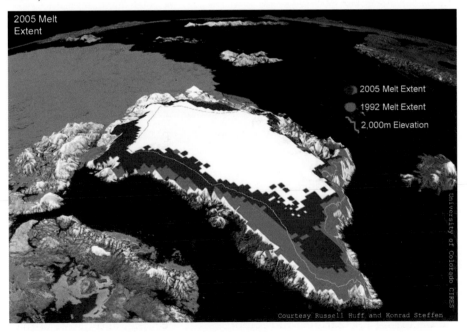

FIGURE 10.8 This 3-D map shows the extent of melting of the Greenland ice sheet between 1992 and 2005. The pink areas are the melt regions in 1992 and the red areas are the new melt regions for 2005. The 2,000-meter elevation is shown by the thin white line. (Courtesy Russell Huff and Konrad Stoffen, CIRES, University of Colorado, Boulder, CO)

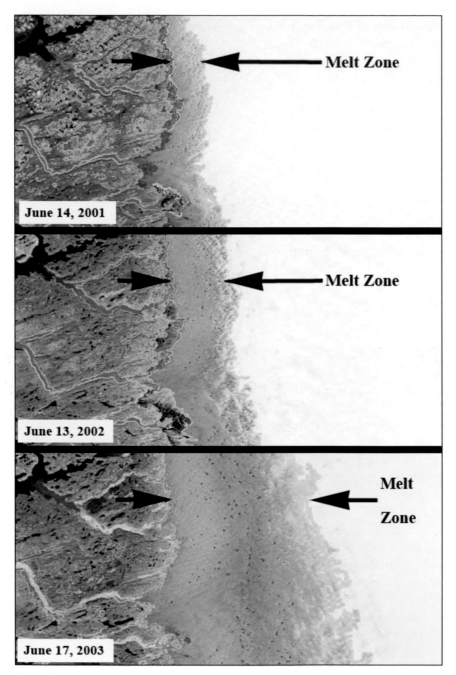

FIGURE 10.9 These satellite images show the increasing extent of melting on one part of the Greenland ice sheet between 2001 and 2003. Each frame is about 150 km across. (Modified from Jacques Descloitres, MODIS Land Rapid Response Team, NASA Goddard Space Flight Center, Greenbelt, Md.)

About half of this was from greater summer melting, and the rest from increased glacial speeds that, in some cases, exceeded those needed to balance the upstream snow accumulations. One large glacier's velocity almost doubled between 1997 and 2003 due to increased melting and loss of ice tongue barriers. This resulted in a net loss of ice from its drainage basin of about 20 km^3 of ice between 2002 and 2003.

In the first direct, comprehensive mass survey of the entire Greenland ice sheet, scientists using data from the NASA/German Aerospace Center Gravity Recovery and Climate Experiment (GRACE) have measured a significant decrease in the mass of the Greenland ice sheet. It decreased by 239 $\pm$ 23 km^3/year between 2002 and 2005 (Chen et al., 2006; Velicogna and Wahr, 2006), which is higher than all earlier estimates and agrees with satellite radar data. The loss of mass more than doubled from 4–50 billion tons per year between 1994 and 1999 to 57–105 billion tons between 1999 and 2004 (Thomas et al., 2006). This amount of melting represents a rise in sea level of about 0.5 mm per year.

Figure 10.10 shows meltwater due to summer melting pouring down a hole (called a moulin) in the ice sheet. This consequence of melting is destabilizing the ice sheet margins by causing them to slide more rapidly toward the ocean (Zwally et al., 2002). The rapid migration of surface meltwater to the ice–bedrock interface is partly to blame for an acceleration of portions of the ice sheet toward the ocean. The variations in the accelerations are correlated with variations in the intensity of surface melting; larger accelerations are associated with higher amounts of surface melting. Furthermore, the movement of the ice sheet takes place all year long. The water at the base of the glacier does not freeze because it is at its pressure-melting point of $-1\,°C$. As meltwater accumulates at the base of the glacier, the ice moves faster and faster with the greatest acceleration coming in the summer melt season. This accounts for about 10% of the increase in glacial acceleration. The primary reason is the collapse of the ice shelves that caused a barrier to glacial movement.

The Jakobshavn Isbrae glacier is Greenland's largest outlet glacier draining 6.5% of the interior ice sheet area (Figure 10.11). It slowed down from 6.7 km/year in 1985 to 5.7 km/year in 1992, but by 2000 movement had increased to 9.4 km/year and in 2003 it had increased to 12.6 km/year (Joughin et al., 2004). At least part of the increase between 1992 and 2003 was probably due to increased melting and migration of the meltwater to the base of the glacier. However, most of the acceleration is probably due to the loss of the floating ice tongue that was restraining glacial motion. New measurements (Rignot and Kanagaratnam, 2006) show that the frontal speed has increased 95% during the progressive break-up of its floating ice tongue.

FIGURE 10.10 Meltwater on Greenland's ice sheet is pouring down a hole in the ice (a moulin) to eventually reach the base of the ice sheet causing it to move more rapidly to the ocean. (Source: Roger Braithwaite, University of Manchester, UK)

The melting of Greenland ice has increased dramatically during the summer of 2005. New measurements show a sudden decline in the very large Helheim glacier on the east side of Greenland (Howat et al., 2005). This large glacier extends from the inland ice cap to the sea through a narrow gap in a mountain range on the island's east coast. The glacier has lowered 30 meters during the summer of 2005 and the front of the glacier has retreated 7 kilometers during the last four years. Also, the

FIGURE 10.11 *This view of the Jakobshavn Isbrae glacier outlet in Greenland shows its break-up and dumping of icebergs into the ocean raising sea level. (Source: Professor Konrad Stoffen, University of Colorado, Boulder, CO)*

Kangerdlugssuaq glacier has begun to move at more rapid rates. Within the last two years (2004–2006) these glaciers in east Greenland have experienced accelerations and retreats similar to the Jakohshavn Isbrae glacier in west Greenland (Luckman et al., 2006). This has resulted in a doubling of the ice flux to the ocean from this part of Greenland from about 50 to 100 km^3/year. These three glaciers, acting in a similar way after a long period of stability, suggests that the Greenland ice sheet is rapidly becoming unstable.

Confirmation of this increasing instability has come from a study of earthquakes from moving Greenland glaciers (Ekström et al., 2006). Periodically glaciers will lurch forward with enough force to generate elastic seismic waves that are recorded on the world's seismometers. On

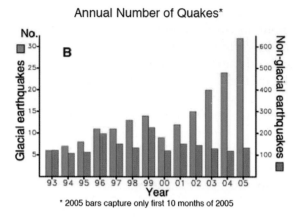

FIGURE 10.12. *This diagram shows the increase of glacial earthquakes caused by movement of portions of the Greenland ice sheet. The rapid increase indicates that the instability of the ice sheet is increasing, and that there is a danger of rapid collapses raising sea level in short periods of time. (Source: Ekstrom, Nettles and Tsai, Science, 311, 1756, 2006)*

Greenland these glacial earthquakes show a strong correlation with the seasons and a large recent increase (Figure 10.12). They are much more frequent in the summer when melting is greater, with the meltwater lubricating the base of glaciers. During the past five years their rate of occurrence has doubled, indicating the growing instability of the ice sheet. The magnitude of the quakes (4.6–5.1) indicate movements of about 10 km^3 of ice to distances of about 10 meters over a duration of 0.5 to 1 minute.

Recent analysis from the data sets of three satellites acquired between 1996 and 2005, and covering 75% of Greenland, have provided some very disturbing information (Rignot and Kanagaratnam, 2006). The net ice loss in 1996 was 96 km^3/year, but had doubled to 224 km^3/year by 2005. This strongly indicates that the ice sheet has become unstable due to melting and sea ice destruction. This suggests that the sea level will rise much faster than anticipated by previous models. Greenland's mass loss doubled in the last decade, contributing to an increased sea level rise from 0.23 mm/year in 1996 to 0.57 mm/year in 2005. In fact, as global temperatures increase (high northern latitudes are increasing fastest), a catastrophic sea level rise in the near future is not out of the question. Ultimately the melting and break-up of the Greenland ice sheet is due to global warming. Temperature records at Angmassalik, Greenland, at 65.6 °N latitude show a 3 °C increase in yearly air temperatures between 1981 and 2005. It is possible that the Greenland ice sheet is entering a state of collapse (Christoffersen and Hambrey, 2006).

The melting of sea ice does not raise the sea level because it displaces about an equal amount of water. However, this is not the case for ice on land. When it enters the ocean it raises sea level. (If you fill a glass with ice and add water to the top of the glass, water will not overflow the glass when the ice melts because the ice will displace an equal amount of water. If you fill a glass with water up to its brim and then add ice, the water will overflow the glass because the new ice displaces the water.) Sea level will rise as the Greenland ice sheet slowly advances into the ocean. So there are two ways of raising sea level from ice. One is the addition of meltwater itself to the oceans, and the other is the addition of large amounts of land ice to the oceans. The addition of meltwater is the slowest. Computer models indicate that if the Earth's atmosphere rose 3 °C the meltwater alone from the Greenland ice sheet would raise the sea level about 7 meters in 1,000 years (Gregory et al., 2004). These models only take into account melting, and that is not the most important way that glaciers disintegrate. If there would be large collapses of land glaciers into the sea, which is beginning to happen in Greenland, then the sea level will rise much faster.

A study of the last interglacial period 130,000 years ago (Eemian Interglacial) suggests we may be underestimating the amount of sea level rise ahead of us (Overpeck et al., 2006; Otto-Bliesner et al., 2006). During this interglacial period the temperature was about 3° to 5 °C warmer than today in the high northern latitudes because of the Earth's different eccentricity (Kaspar et al., 2005) and a slightly greater axial tilt. At that time the melting and collapse of the Greenland ice sheet and other circum-Artic ice fields was much more extensive than today and contributed a 2.2 to 3.4 meter rise in sea level. Computer models suggest that by 2100 the temperature will be as high, resulting in similar melting and collapse of the Greenland ice sheet that would produce a similar rise in sea level.

ANTARCTICA

The largest ice mass on Earth is the Antarctic ice sheet. It comprises 90% of all ice on Earth, and, if melted, would raise the sea level about 60 meters. The interior eastern part of this ice sheet has actually gotten cooler and accumulated more ice in the past 25 years. The reason for the cooling is probably the concentration of cold air near the center caused by the ozone hole (Shindell and Schmidt, 2004). The accumulation of ice (about 45 billion metric tons) is probably the result of global warming causing more ocean evaporation and, hence, more precipitation in east Antarctica (Davis et al., 2005). However, recent satellite measurements from the GRACE mission between 2002 and 2005 show that there has been a significant decrease in the ice sheet's mass (Velicogna and Wahr, 2005 and 2006). There has been a decrease in the ice sheet of about 152 km^3 each year between 2002 and 2005; enough to raise the sea level by 1.2 mm. Most of this mass loss is from the west Antarctic ice sheet. Therefore, both the Greenland and Antarctic ice sheets are rapidly losing their ice.

Recent radiosonde observations indicate that the mid-troposphere over Antarctica has warmed faster than anywhere else on Earth. Temperatures have increased at a rate of 0.5 to 0.7 °C per decade over the past 30 years (Turner el al., 2006), although similar temperature increases have not been observed at the surface. The cause of this warming is not yet known, but increases in greenhouse gases are probably playing a role. The surface of west Antarctica and the Antarctic Peninsula are warming, resulting in melting and increased movement of glaciers into the sea.

Sea Ice

The ice shelves of the Antarctic Peninsula and west Antarctica are thinning and beginning to collapse. Sediment cores indicate that there were fewer ice shelves along the Antarctic Peninsula between about 9,000 and 2,500 years ago. This is thought to be due to what is called the Antarctic Oscillation, a high-pressure ring of warmer air that used to wander, but now stays mostly in one location due to the ozone hole. In recent times the sea ice of the eastern side of the Antarctic Peninsula has begun to collapse due to melting. The east Antarctic Peninsula sea ice is called the Larsen ice shelf. It is divided into Larsen A, B, and C. Larsen A collapsed in the 1990s, and Larsen B has started collapsing. Larsen C is rapidly melting and will soon start to collapse. The record of diatoms (small marine animals), detrital material, and geochemical parameters from sediment cores in the vicinity of the Larsen B ice shelf indicate that the recent collapse is unprecedented during the past 12,000 years (Domack et al., 2005). This indicates that the present collapses are probably due to global warming. The Antarctic Peninsula is warming at a rate of about 0.5 °C per decade. It has warmed about 3 °C since 1951, primarily during the Southern Hemisphere fall and winter (King and Comiso, 2003; Skvarca et al., 1999; Turner et al., 2005). This is probably primarily due to global warming, but aided by warm air concentration from circumpolar westerly winds on the Antarctic Peninsula mountains (Orr et al., 2004). The adjacent ocean had temperature and salinity changes at the same time with summer temperatures rising more than $1\,^{\circ}C$ and strong upper-layer increases in salt content (Meredith and King, 2005). These changes will have positive feedbacks that will contribute to continuing climate change.

On average the Antarctic Peninsula ice shelves have retreated by about 300 km^2/year since 1980 (Vaughan and Doake, 1996). Satellite data show that the Larsen ice shelf lowered by about 0.3 meter/year between 1992 and 2001 due to increased summer meltwater and the loss of basal ice through melting from warming ocean waters (Shepherd et al., 2004). This gradual retreat was accompanied by two catastrophic collapses in January 1995 and February 2002. Figure 10.13 shows satellite views of the 2002 collapse that took place within a period of three weeks during the Antarctic summer. At this time melting was three times greater than the previous five summers (van den Broeke, 2005). The collapse resulted in a 3,250 km^2 loss of the ice shelf, at which time the shelf was 220 meters thick. This ice shelf has lost 60% of its area in the last five years, and over 13,500 km^2 of all Antarctic Peninsula ice shelves have been lost since 1974.

At one time it was debated whether or not ice shelves acted as a barrier for the advance of continental glaciers toward the sea. We now know that, in fact, they are barriers (Dupont and Alley, 2005). As a consequence of the lost ice shelves, the rate of ice flow of the continental glaciers toward the ocean increased two-fold in early 2003 and three-fold by the end of 2003 (Rignot et al., 2004; Scambos et al., 2004; De Angelis and Skvarca, 2003). The loss of mass into the sea associated with the accelerated flow exceeds 27 km^3/year, and the ice is thinning at a rate of tens of meters per year.

On the western side of the Antarctic Peninsula the Wordie Bay ice shelf collapsed in 1974, and other collapses occurred in 1989 and 1998. New satellite data indicate that glaciers now flowing into Wordie Bay discharge about 6.8 km^3/year of ice into the bay (Rignot et al., 2005). This is 84% larger than the snow accumulation over the 6,300 km^2 drainage basin. Glaciers are thinning at a rate of 2 meters/year. Furthermore, Fleming Glacier flows 50% faster than it did before the ice shelf collapse in 1974.

Other glaciers along the Amundsen Coast of the west Antarctic ice sheet are discharging ice in excess of accumulation. Ice shelf thinning rates of Pine Island Bay were about 5.5 meters/year during the past decade. The cause of this thinning is probably ocean currents that are on average $0.5\,^{\circ}C$ above freezing (Shepherd et al., 2003). In response, large glaciers feeding the Amundsen Coast ice shelves have thinned and accelerated by up to 26% over the last three decades, with perturbations extending more than

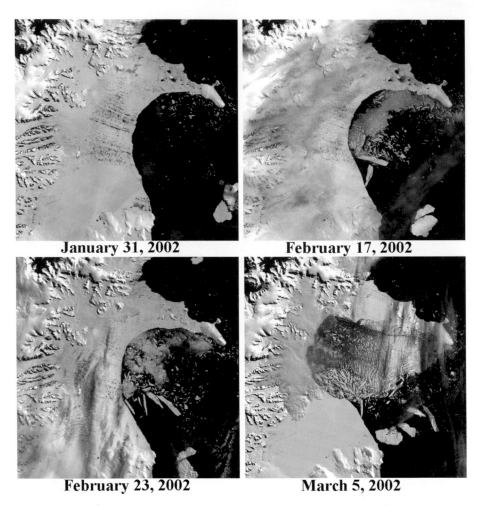

January 31, 2002   **February 17, 2002**

February 23, 2002   **March 5, 2002**

FIGURE 10.13 This series of satellite images shows the break-up of the Larsen B ice shelf on the Antarctic Peninsula during a three-week summer period in 2002. In the January 31 image the dark rows are meltwater lakes on the ice shelf that were instrumental in the collapse. Similar meltwater lakes are forming in the summer on the Greenland ice sheet. (Modified from Ted Scambos, National Snow and Ice Center, Univ. of Colorado, MODIS)

200 km inland (Shepherd et al., 2003; Thomas et al., 2004; Joughin et al., 2003; Payne et al., 2004). This and other circumstances are beginning to have a destabilizing effect on the west Antarctic ice sheet.

West Antarctic Ice Sheet

It is possible that the west Antarctic ice sheet disintegrated during the last interglacial period some 125,000 years ago without the addition of greenhouse gases caused by fossil fuel burning (Bentley, 1980). Today there is a possibility of another collapse due to global warming. Much of west Antarctica is a "marine ice sheet" that has its bed well below sea level. This is potentially an unstable condition. The ice streams that make up this portion of the ice sheet have their beds resting on and lubricated by muddy sediments. This means that only small changes in the distribution and temperature of the water at their base can have very large effects on the velocity of the ice.

Until recently, it was considered unlikely that the west Antarctic ice sheet would collapse before 2100. This whole idea has to be reconsidered because of new findings. The movement of most west Antarctic glaciers is accelerating. West Antarctica was thought to be losing as much as 76 km^3/year to the sea. The total loss was thought to be enough to raise sea level about 0.16 mm/year. The temperature of the Antarctic Ocean has risen about 1 °C since the 1960s. Ice shelves in the Amundsen Sea Embayment and the Pine Island Bay are thinning rapidly due to basal melting from warming seas that are now 0.5 °C above freezing (Shepherd et al., 2004). This is allowing the glaciers to move more rapidly into the sea. Recent measurements (2005) by the British Antarctic Survey of the ice discharge rate into Amundsen Bay indicates that ice is flowing into the bay at a rate of about 250 km^3/year. This was only recently discovered because the area is so remote and inaccessible, and if this rate continues, it alone is enough to raise sea level by 0.2 mm/year. This high discharge rate could indicate that the ice sheet is becoming increasingly unstable.

If the Pine Island glacier that drains most of west Antarctica retreated enough to produce a hole in the side of the ice sheet so that the remaining ice drained through it, a collapse could be imminent (Alley, 1998). Once enough ice had drained through the hole, most of the west Antarctic ice sheet would eventually collapse. This would raise sea level about 6 meters. Such an increase would permanently flood about 30% of Florida and Louisiana, 20% of Maryland and Delaware including Washington, DC, and about 10% of the Carolinas and New Jersey. Obviously, the major port cities of the world would be under water and commerce would be catastrophically disrupted.

Probably the most important finding about ice in the past 20 years is the rapidity with which substantial changes can occur on polar ice sheets. This should serve as a warning that we cannot be complacent about global warming. If a substantial collapse of the Greenland or west Antarctic ice sheets should occur, it could raise sea level 6 meters over a few centuries and be an unparalleled disaster for humankind.

PERMAFROST AND GROUND ICE

One of the scariest aspects of global warming is that permafrost is beginning to melt (Figure 10.14). Permafrost absorbs about 14% of the carbon emitted by sources. Most of the world's permafrost is located in northern Russia, the rest occurs primarily in Alaska, Canada, and Scandinavia. Most of this permafrost is now beginning to melt because of global warming. As the permafrost melts it turns into lakes, and as melting proceeds the lakes drain into the melted subsurface leaving dry lakebeds. Excluding ice sheets, the area underlain by permafrost is about 10.5 million km^2, but computer models suggest that near-surface deposits could decrease to 1 million km^2 by 2100 (Figure 10.15), leading to huge releases of methane and CO_2 (Lawrence and Slater, 2005; Stendel and Christensen, 2002).

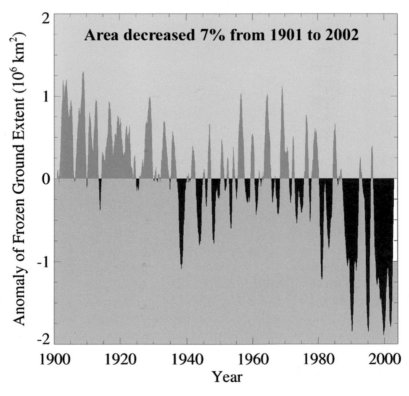

FIGURE 10.14 This diagram shows the decrease in the extent of permafrost from 1901 to 2002. During the last 15 years the decrease has been particularly severe. (From UN IPCC 4th Assessment 2007)

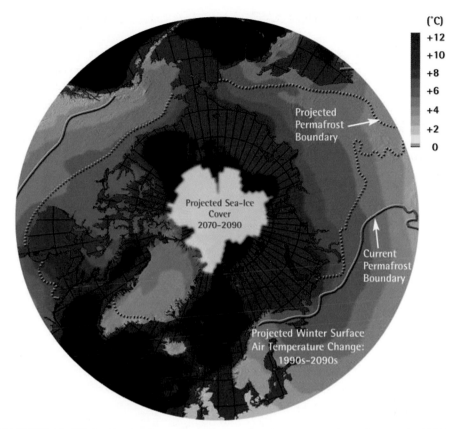

FIGURE 10.15 Map showing the projected winter surface air temperature change between the 1990s and 2090s with the current permafrost boundary and the projected boundary near the end of the century. (From the Arctic Climate Impact Assessment Report, 2004)

In Siberia and central Alaska a large component of the permafrost consists of a glacial wind-blown sediment called *loess* (Zimov et al., 2006). This permafrost contains between 2% and 5% carbon—that is about 10 to 30 times the carbon found in deep non-permafrost soils. It is estimated that ~ 450 billion tons of carbon is locked up in this type of permafrost (Zimov et al., 2006). There is, in addition, ~ 400 billion tons in non-loess permafrost. Experiments indicate that when initially thawed, the permafrost emits 10–40 grams of carbon/m^3 each day and then 0.5–5 grams/m^3 each day over several years. At these rates all of the carbon would be lost in about a century, adding at least an additional 900 billion tons of carbon to the atmosphere.

Western Siberia has warmed faster than almost anywhere else on the planet. Here the temperature has increased about 3 °C over the past 40

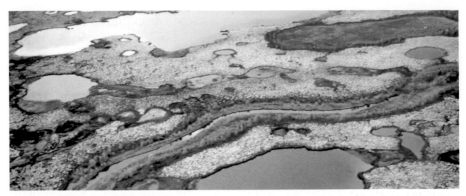

FIGURE 10.16 *Air photo of a portion of the west Siberian plains showing lakes formed by melting permafrost. (Photo by Sergey N. Kirptin)*

years. As a consequence, the world's largest frozen peat bog that formed after the last ice age has begun to melt for the first time (Pearce, 2005). The area that is melting is over a million km^2, or about the combined areas of France and Germany, and is turning into a vast number of shallow lakes (Figure 10.16). A comparison of satellite images taken in the early 1970s with those taken between 1997 and 2004 show a strong decline in lake abundance and area (Smith et al., 2005). The number of large lakes declined by 11% and the lake surface area decreased by 6%. This sharp decline is due to advanced thawing leading to drainage to the subsurface. It is estimated that the west Siberia permafrost contains about 70 billion metric tons of methane, which is 25% of all methane stored on the world's land surface. As the bogs dry out the methane is oxidized and released to the atmosphere as CO_2. However, if the bog remains wet, as most of them are today, the methane is released to the atmosphere before it oxidizes. As a greenhouse gas, methane is 23 times more potent than CO_2. Areas in eastern Siberia have been found where methane is bubbling up from the thawing permafrost so fast that it prevents the surface water from freezing in the middle of winter. Eventually this will surely lead to an acceleration of global warming.

A study of the emission of methane from melting permafrost in northern Siberia shows it is 58% greater than previously thought (Walter et al., 2006). It is estimated that northern Siberia is now emitting about 3.8 million metric tons of methane each year. This is equivalent to the warming potential of about 235 million metric tons of CO_2 each year over a 20-year period. The study does not take into consideration the emission of methane from other arctic regions such as Alaska, Canada, and Scandinavia. It is probable that emissions are considerably more when these areas are taken into account. Although this emission rate is small compared to human-caused emissions of CO_2 ($\sim$ 25 billion metric tons), it indicates a positive feedback to global warming from the growing rate of emissions as Arctic permafrost warms.

It is not only the Siberian permafrost that is beginning to emit methane. Its wetlands have now also become a source of methane to the atmosphere (Friborg et al., 2003). A recent study shows that although the wetlands are still a large sink for CO_2, they are now a source of methane because of global warming. The estimated yearly emission of methane is 26 mg/m^2 (milligrams per square meter). Over the 20- and 100-year time horizons the total greenhouse gas emissions each year from this huge area of wetlands is about 230 and 1,400 million metric tons of CO_2 equivalent (methane oxidizes to carbon dioxide). This is about 6% of the world total CO_2 emission from burning fossil fuels (Friborg et al., 2003). Computer simulations show that methane emissions from wetlands will increase dramatically as the Earth warms (Gedney et al., 2004; Shindell et al., 2004). One simulation with a doubling of atmospheric CO_2 suggests that the Earth's wetlands could release by 27.7 billion metric tons of methane per year or a 78% rise over present emissions (Shindell et al., 2004). Furthermore, wetlands at high northern latitudes expand and emissions nearly triple during the northern summer. The global increase in methane represents about a 20% increase over present values.

Siberia is not the only place in which permafrost is thawing. In Alaska, Canada, and Scandinavia major thawing of permafrost is taking place and all of these areas are beginning to release methane and carbon dioxide. Even abrupt large increases of permafrost degradation have occurred in Arctic Alaska due to record warm temperatures during 1989–1998 (Jorgenson et al., 2006). On the North Slope of Alaska fringing the Arctic Ocean there has been a major expansion of lakes indicating large-scale permafrost melting is advancing. A study of over 1.7 million km^2 of Canadian peatlands in 2001 indicates that 235 million metric tons of CO_2 equivalent are released to the atmosphere each year primarily from fires (Turetsky et al., 2002). These disturbances have reduced the carbon uptake by 85%. A 17% increase in the area of peatland burning will convert these peatlands from a sink to a net source of carbon to the atmosphere. Studies of the permafrost bogs in sub-arctic Sweden show that global warming has thawed much of it and changed the vegetation so that methane is now being emitted. The past mean annual temperature in the study area was –0.7 °C, but it has risen above 0 °C in recent years because of global warming. For the period 1970–2000, the increase in methane emission ranges from 22% to 66% (Christensen et al., 2004).

In addition to releasing greenhouse gases the thawing permafrost is

FIGURE 10.17 A portion of the foundation of this apartment building in Cherskii, eastern Siberia, collapsed from the melting permafrost on which it was built. (Reproduced from Goldman, E. 2002. Even in the high Arctic nothing is permanent. Science 297, 1493–1494)

destabilizing the surface so that structures are beginning to crumble. In the arctic regions buildings and other structures that are built on permafrost are becoming unstable as the permafrost is melting (Figure 10.17). In some areas the uppermost layers of the permafrost are thawing at rates close to 20 cm/year. As a result, roads are buckling and cracking, airport runways are fracturing, and buildings are cracking, tilting, and sometimes toppling over. When these structures were built 30 years ago nobody thought about climate-induced changes in ground temperatures. In Siberia, 300 apartment buildings in Norilsk and Yakutsk have been damaged so far. If the permafrost keeps warming at the very modest rate of just 0.075 °C per year, all the five-story buildings in the city of Yakutsh (Population 193,000) will be destroyed by 2030. Similar things are happening in other areas of permafrost around the world. Sustainable development in the High Arctic will require finding ways to adapt to thawing permafrost, including shoring-up existing structures. This will be an expensive proposition.

ARCTIC TUNDRA

Almost one-third of the world's carbon is stored in the northern latitude soils. Over the 20th century the Canadian annual mean soil temperature at a depth of 20 cm increased 0.6 °C while the annual mean air temperature increased 1.0 °C (Zhang and Chen, 2005). The way these soils respond to warming is important in understanding the world carbon balance. It has been suggested that warming soils in tundra and boreal regions will increase decomposition of plant litter, increasing nutrient availability (Hobbie et al., 2002). This, in turn, may stimulate plant production and increase carbon storage which would partially offset increasing atmospheric CO_2 (Shaver et al., 1992). However, a new long-term experiment with Alaskan tundra indicates that this is not the case (Mack et al., 2004). The experiment was a long-term fertilization of tundra that increased the nutrient availability causing a net loss of 2 kg of carbon/m^2 over 20 years. The plant production doubled during the experiment. However, the losses of carbon and nitrogen to the atmosphere from the deep soil layers more than offset the increased carbon and nitrogen storage in plants and litter. This study suggests that warmer temperatures due to global warming will increase soil nutrients and amplify carbon release to the atmosphere causing a net loss of soil carbon and a positive feedback to global warming.

A study in England and Wales shows that the carbon loss from the soils between 1978 and 2003 due to global warming is offsetting carbon absorption by terrestrial sinks (Bellamy et al., 2005). Probably this is also happening in other temperate regions, causing a positive feedback.

NORTHERN HEMISPHERE SNOW COVER

With regard to land coverage, seasonal snow cover is the largest part of the Earth's cryosphere. It has a winter maximum extent of 47 million km^2, and about 98% of this snow cover is located in the Northern Hemisphere. Snow cover is important to the climate system. Snow reflects about 80 to 90% of the sunlight that falls on it; therefore, a large amount of this sunlight is reflected back to space without heating the surface, and without snow cover the surface would absorb about 4 to 6 times more of the Sun's energy. Global warming, however, is causing the snow cover to decrease.

Satellite data taken over the past 24 years indicates that the average snow cover has decreased about 3 to 5% per decade, particularly in the spring and summer (The National Snow and Ice Data Center, 2005). Because the precipitation in regions of seasonal snow cover appears to have stayed the same or increased slightly over the time period, the diminishing snow cover is probably the result of increasing temperatures. Large amounts of energy are now directly warming the ground where normally there would be snow and this is causing a positive feedback to global warming. Because of the decrease in reflected energy, the absorption of solar radiation increases causing the surface temperatures to rise, causing more snow to melt, causing more heating of the surface, and so on.

River and lake ice has been forming later in winter and breaking up earlier in the spring. On average, the freeze date has become later by about six days per century, and the break-up date about six days earlier per century.

SUMMARY

If you had any doubts that global warming was taking place and was a serious threat, they should have been dispelled by the contents of this chapter. Mountain glaciers are melting at an accelerating rate. The ice shelves at the Arctic and particularly along the Antarctic Peninsula are rapidly melting and collapsing, unplugging the cork that retarded the movement of land glaciers into the sea. Both the Greenland and west Antarctic ice sheets are melting and becoming unstable. The movement of glaciers that drain them is accelerating, and they are dumping large volumes of ice in the oceans, raising the sea level. The Arctic Ocean summer ice shelf is decreasing rapidly. It is now possible that the Arctic has entered an irreversible phase of warming that will rapidly accelerate the loss of the Arctic sea ice. The summer ice shelf may be gone in less than 50 years. Even the winter ice shelf is shrinking. The world's permafrost has begun to melt, and the thawing is accelerating at a rapid rate releasing methane and CO_2 to the atmosphere. Eventually, warming of tundra soils will release more carbon than it stores, causing a positive feedback. Snow cover is diminishing, causing additional heating and a positive feedback. We are in trouble.

11

WATER WORLD

OCEANS cover about 70% of the Earth's surface. No other planet or satellite in the Solar System has liquid water on its surface at the present time. The Earth's oceans have a profound effect on climate, because water has some very unique properties. Water can dissolve more substances in greater amounts than any other liquid. It has a high specific heat, which means that it takes a large number of calories to raise its temperature (1 cal/gram/°C). This high heat capacity allows oceans and large lakes to moderate the temperature of adjacent land, regulating its climate. Water also has the highest heat of vaporization of any liquid (540 calories/gram), which means that large amounts of heat are required to evaporate even small amounts of water. It is this energy release that drives the Earth's weather. Water also has a large thermal inertia, which means that it takes longer to heat, then cool, than many other substances. For instance, when you place a pot of water on a heat source, it does not boil immediately; it takes several minutes. After you remove the heat the water takes many minutes to return to its original temperature. In other words, water gains and loses its heat slowly (it has a large thermal inertia).

The Earth's oceans also transport enormous amounts of heat to various parts of the world. For example, the Gulf Stream each day transports twice as much heat as would be produced by burning all the coal mined globally in a year. It is this transported heat flowing up the North Atlantic Ocean that keeps Europe temperate and livable.

The world's oceans have a profound influence on the Earth's climate, weather, and biology. If the oceans are changed by our actions then there will be enormous effects on all other systems with significant deleterious effects on all of us.

EARTH'S ENERGY IMBALANCE

As stated above, the Earth's climate system, particularly the ocean, has a large thermal inertia. This inertia delays the Earth's response to climate forcings, such as greenhouse gas warming. In other words, the warming of Earth by greenhouse gases is delayed, making future temperature increases inevitable. Figure 11.1 is a diagrammatic representation of how the climate system thermal inertia delays atmospheric CO_2 stability, temperature, and sea level rise due to the melting of ice and the thermal expansion of sea water, even when our CO_2 emissions peak and then dramatically decline. Because the climate sensitivity is about $0.75 \pm 0.25\,°C$ per W/m^2, it takes about 25 to 50 years for the Earth's surface temperature to reach 60% of its equilibrium value (Hansen et al., 2005). However, the greater the sensitivity, the shorter the time. If the sensitivity was $0.25\,°C$ per W/m^2 the time would be only about 10 years, and if it was $1\,°C$ per W/m^2 it would be about 100 years. This delay provides an opportunity to reduce the intensity of global warming before it reaches its maximum, but only if we begin now to significantly reduce greenhouse gas emissions. For this action to be effective it is necessary to eliminate the energy imbalance discussed below. Given the positive feedbacks discussed in Chapter 10, this will be a very difficult, if not impossible, task. However, if we continue to wait for even more overwhelming evidence of global warming and its consequences, the large inertia means that we will unavoidably suffer even greater global warming no matter what we do.

Recent studies show that Earth is absorbing about $0.85\ W/m^2$ more energy than it is radiating back to space (Hansen et al., 2005). Furthermore, from 1993 to 2003 most of that excess energy has been absorbed in the oceans' top 750 meters (Figure 11.2). Consequently, its

CO₂ concentration, temperature, and sea level continue to rise long after emissions are reduced

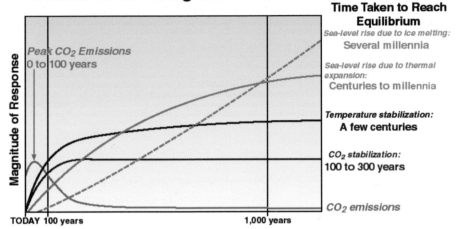

FIGURE 11.1 *This diagrammatic representation shows the period of time before part of the climate system reaches equilibrium long after CO₂ emissions are reduced. This delay is due to the inertia of the climate system, primarily the oceans. Even if we now reduce greenhouse gas emission to zero, we will still have a 0.6 to 1 °C rise in average global temperature. See text for explanation. (From UN IPCC report, 2001)*

full effect on the climate system has yet to be realized. The decade of observations show that Earth's oceans absorbed an average of six excess watt-years of energy per square meter.[1] Although this may sound small, it is large compared to other epochs in Earth history. For example, if that level was maintained throughout the Holocene (the last 10,000 years), it would be enough to raise the temperature of the ocean's top 750 meters by more than 100 °C. This demonstrates that the current imbalance is unprecedented in the last 10,000 years. Furthermore, the ocean temperature would have increased even more had it not been for the cooling effect of aerosols (Delworth et al., 2005)

The increase in the average global temperature of 0.5 °C since 1980 is not enough to account for the total energy imbalance. Of course some of the excess heat has melted ice and warmed the land surface, but most of it has been stored in the ocean since 1980 and will make its presence felt in the future (see Table 6.1 in Chapter 6). Because of the warmer oceans, we can expect an additional 0.6 °C increase in average global temperatures even if we stopped increasing greenhouse gas concentrations because there is a "lag time" between the ocean's absorption of the excess energy and

---

[1] A watt-year is the total amount of energy supplied by 1 watt of power for a year.

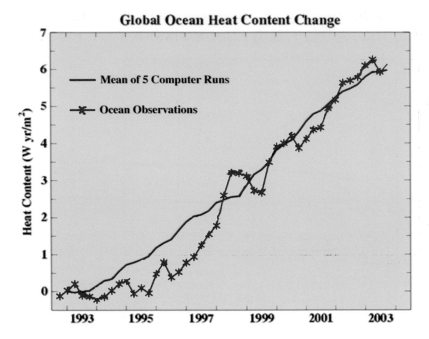

Global Ocean Heat Content Change

FIGURE 11.2 This graph shows the increase in the ocean's heat content from buoy observations together with the mean of 5 computer runs to simulate this rise using human-caused greenhouse gases. See text for explanation. (From Hansen et al., 2005, with modifications)

when that excess produces changes in the global average temperature (Hansen et al., 2005). We must realize that what we have already done has repercussions far into the future. Several studies suggest that even if the radiative forcing is fixed at today's levels the global mean surface air temperature will rise another 0.6 to 1 °C before it reaches equilibrium (Wetherald et al., 2001; Meehl et al., 2005; Wigley et al., 2005; Hansen et al., 2005). In other words, we have already "shot ourselves in the foot." Our immediate problem is not to "shoot ourselves in the head."

THE CHANGING OCEANS

Global warming is beginning to change the world's oceans in profound ways—in particular, the temperature, chemistry, sea level, and biology of the oceans are changing in response to self-inflicted global warming. Because of the oceans' large thermal inertia, changes related to temperature will persist for hundreds of years no matter what we do.

Temperature

In 1965 it was first discovered that the oceans were warming. Since then the oceans have warmed at an alarming rate (Hegerl and Bindoff, 2005). Even the deep frigid bottom waters have warmed; for example, in the North Atlantic Ocean temperatures reached an all-time high in 2004. The ocean surface about 7 km off St Johns, Newfoundland, averaged 1.1 °C above normal, the highest ever recorded. Also the bottom temperature at a depth of 176 meters was 0.9 °C above normal.

However, the oceans, like the atmosphere, are not warming uniformly. One reason is that cooling aerosols produced by nature and humans are not uniformly distributed in the atmosphere, and, therefore, do not result in a uniform decrease in absorbed surface solar radiation. For example, there was decreased warming of the Northern Indian Ocean due to the abundance of aerosols produced in that part of the world. Another reason is that changes in the Earth's radiation balance can cause regional changes in atmospheric and ocean circulation that affect the amount of heating on a regional scale.

Figure 11.3 shows the change in the average temperature anomalies for the world's oceans from about 1860 to 2000. Although variable, the period after 1970 shows a strong warming trend in most of the world's

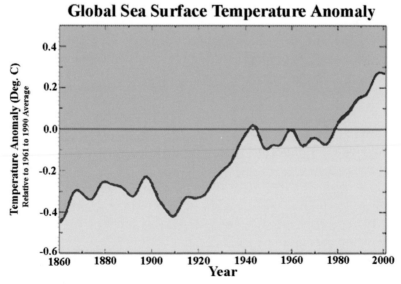

FIGURE 11.3 This graph shows the change in average temperature anomalies for the world's oceans from about 1860 to 2000. The gain in temperature after 1975 coincides with the increasing atmospheric surface temperatures since that time. Modified from the IPCC 2001 report and data of UK Met Office. (From Jones et al., 2001)

oceans. This corresponds to the surface air temperature rise during the same period (see Figure 6.5). During the past 20 years the Indian Ocean and the southwest Pacific Ocean warmed 0.5 °C. At depths of 200 to 1,000 meters in the Arctic Ocean it was discovered that the water had warmed 1 °C in 10 years, and the Bering Sea depth-averaged temperature was 2 °C warmer in 2003 than the mean temperature between 1995 and 1997. The Atlantic Ocean in the region 24.5 °N is warming at a rate of 0.009 °C per year at a depth of 1.1 km, and 0.042 °C per year at a depth of 400 meters (Vargas-Yáñez et al., 2004). For the world's oceans from 0 to 3 km depth, the average temperature has increased by 0.037 °C during the period 1955 to 1998 (Levitus et al., 2005). This may sound like a very small amount, but as water has a large heat capacity it is roughly equivalent to the heat stored by the oceans resulting from seasonal heating in a typical year, and that is a large number. Increases in the temperature of the oceans, in turn, affect the sea levels, chemistry, biology, and local climate.

The absorption of CO_2 by ocean water depends on its temperature; the warmer the water the less CO_2 is absorbed. For each 1 °C the water warms, it absorbs 4% less CO_2. The oceans have warmed on average about 1 °C during the last 50 years, so the oceans are taking up about 4% less CO_2 solely due to this small rise in temperature. The amount of dissolved oxygen is also decreasing due to changes in the circulation by warming oceans. This has led to a decline in the uptake of CO_2 by about 2.5 billion metric tons per year (Joos et al., 2003). The uptake of CO_2 in the North Atlantic has decreased, probably due to a decrease in biological activity in the region caused by the warming oceans (Lefevre et al., 2004).

Sea Level

There is a correlation between sea level and the CO_2 content of the atmosphere (Figure 11.4). Greater than 35 million years ago the CO_2 content was about 1,250 ppm and the average global temperature was about 12 °C above today's value. At that time sea level was about 73 meters higher than today because there was much less ice than there is today. About 32 million years ago the CO_2 content decreased to about 500 ppm and the sea level was about 45 meters higher than today. This was a time when the Antarctic ice sheet first began to grow. About 21,000 years ago, during the last glacial maximum, the CO_2 content was only 185 ppm and sea level was 130 meters lower than today. During these times the CO_2 content must have significantly regulated the global temperature, and, thus, the amount of ice and sea level—the less ice, the higher the sea level. The current sea level rise also correlates with rising atmospheric CO_2.

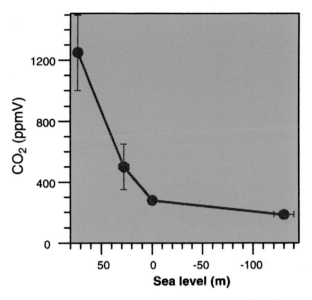

FIGURE 11.4 This graph shows the correlation of sea level (no thermal expansion) with atmospheric CO_2 content from various climate proxies. Zero is sea level with 278 ppm CO_2. The correlation between the rise in sea level and CO_2 is because of temperature, and melting or accumulation of ice. (From Alley et al., 2005)

Since the depths of the last ice age about 19,000 years ago, ∼50 million cubic kilometers of ice have melted from land-based ice sheets raising the sea level 130 meters (see Figure 5.6). Sea level has been steadily rising during the 20th century for an estimated increase of about 20 cm (Figure 11.5). There are two reasons for this rise: (1) new water is being supplied to the oceans by melting ice (Chapter 10), and (2) seawater is expanding as the oceans get warmer. In fact, thermal expansion is thought to account for at least half the observed rate of sea level rise (Antonov et al., 2005).

The sea level is not rising uniformly. The largest sea level rise has been in the western Pacific Ocean and the eastern Indian Ocean. All of the Atlantic Ocean has risen, but the western Indian Ocean and the eastern Pacific Ocean have fallen. The pattern of sea level rise and fall is consistent with the pattern of thermal expansion due to differences in the heating of different parts of the world's oceans, and differences in the salinity, winds, and circulation.

The accelerated discharge of Greenland's outlet glaciers may have contributed up to ∼0.09 mm/year in sea level rise since the mid-1990s (Joughin et al., 2004). The recent disintegration of glaciers along the Amundsen coast of the west Antarctic ice sheet is estimated to have contributed about 0.13 to 0.24 mm/year to the rise in sea level (Shepherd

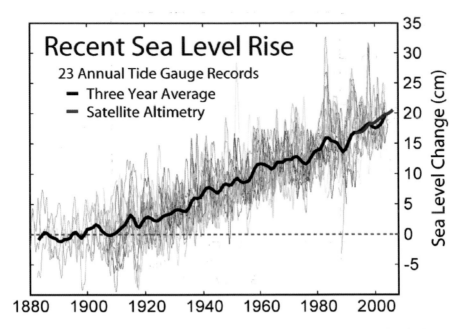

FIGURE 11.5 *The rise in sea level from about 1850 to 2005 is shown in this graph. The data are from tide gauge records. The black line is a three-year average. The red line is from TOPEX/Poseidon satellite data. (From Robert A. Rohde Global Warming Art project)*

et al., 2004; Thomas et al., 2004). However, Greenland presently makes the largest contribution to the rise in the sea level (Alley et al., 2005).

There is still some uncertainty about the current rate of sea level rise. Coastal and island tidal gauge measurements indicate a rate of 1.8 mm/year (White et al., 2005), but the Topex–Poseiden spacecraft high precision sea level measurements give a value of 2.8 mm/year over a 12-year time period (Cazenave and Nerem, 2004). A recent study of sea level changes over the last 100 million years indicates that sea levels have risen twice as fast during the past 150 years than in previous centuries (Oerlemans, 2005). A new study of sea level rise since 1870 indicates that it has risen an average of 1.7 mm/year, with the rate of rise increasing an average of 0.01 mm/year (Church and White, 2006). The highest rate has been since the early 1990s, with sea levels increasing by about 3 mm/year. If this acceleration continues, by 2100 the sea level will be about 31 cm higher, on average, than it was in 1990. Because of the increasingly rapid disintegration of the Greenland ice sheet we may be in for much larger sea level rises in the future. However, even the relatively small rises in sea level are having significant effects on our beaches.

Seventy percent of the world's sandy beaches are noticeably eroding,

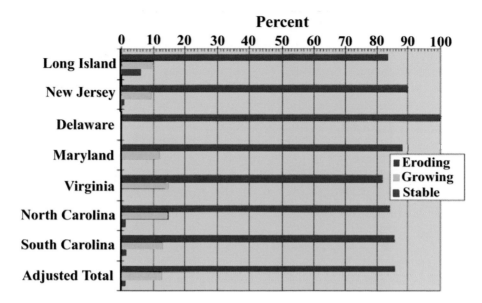

FIGURE 11.6 This bar chart shows the percent of the beaches that are stable, growing and eroding on the eastern seaboard of the United States. Over 80% of all beaches are eroding largely due to rising sea level from global warming. (Modified from Leatherman et al., 2003)

including 90% of the beaches in the United States (Heinz Center, 2000). Figure 11.6 shows the percentage of beaches in erosion, accretion, and stable in various regions of the east coast of the United States (Leatherman et al., 2003). The vast majority of the beaches are eroding, and parts of the Texas Gulf Coast are retreating at a rate of 1.5 meters/year. The rate of landward beach erosion depends on many different processes, including the presence and intensity of longshore currents, the roughness and slope of the beach, etc. However, the sensitivity to erosion from the rate of sea level rise can be estimated from the slope and distance of the berm.[2] This is a highly simplified model that cannot predict the behavior of any individual beach, but it can be used to reveal how sensitive erosion is to sea level rise. For example, a sea level rise of 1.8 mm/year suggests a landward retreat of 9 to 18 cm/year, and for a sea level rise of 2.8 mm/year the retreat is 14 to 28 cm/year (the present rise is 3 mm/year). The latter erosion rate could completely erase many of the world's beaches (depending on other factors) over the next 20 or 30 years unless action is taken to protect them. Furthermore, storm surges will become much more destructive as sea level rises. For example, the most damaging storm surge on the east coast of the

[2] That part of a beach where the sloping, sea-washed sand meets the flat, dry sand higher up.

United States was due to a winter storm that occurred when the water was particularly high on March 5–8, 1962, from what is known as a perigean tide; and the storm surge from hurricane Katrina in 2005 broke the New Orleans levee and flooded the city with catastrophic results.

Over a half-million people in the United States live within 150 meters of the shoreline. A 2000 study estimated that about 1,500 homes and the land on which they are built will be lost to erosion each year over the next several decades (Heinz Center, 2000). Costs to coastal property owners will average $530 million/year. However, this study used a sea level rise that was lower than today's estimate, so this projection is probably too low. Figure 11.7 shows the projected housing loss in South Bethany, Delaware, and a house that has been almost destroyed from beach erosion.

The rising waters are also starting to destroy low-lying island nations. For example, the Tuvalu island nation was formed in 1978 when the British Colonial authorities handed over power to the islanders of half the Gilbert and Ellice Islands. The nine islands are only 1-4 meters above sea level and now they are becoming uninhabitable. Rising sea level, severe coastal erosion, droughts, stronger typhoons, and saltwater intrusion are making the islands uninhabitable. Each year during the "high tide" season, water erupts in gardens, along roads and on to the airstrip (Figure 11.8). Lakes appear in front of doors, and there is nowhere to hide if the tide is accompanied by strong winds or a typhoon. Evacuation of the islands is now required, but a planned evacuation to New Zealand is tied up in "red tape." There was a brief attempt to bring legal action for compensation against the United States and Australia for not ratifying the Kyoto Protocol, but it never went anywhere. As Paani Laupepa, the Tuvaluan head of the environment ministry, put it, "How do you put a price on a whole nation being relocated?" "How do you value a culture that is being wiped out?"

Sea level rise will become a serious problem for the 20% (1.3 billion) of the world's population who live within 30 km of coastal areas. It is projected that the oceans will rise anywhere from 0.5 to 1 meter during the next 100 years. However, the amount of water released into the oceans by melting glaciers and ice sheets may have been underestimated in arriving at the 0.5 to 1 meter figure. Instead of the 1–23 cm rise due to glacier and ice sheet melting used in the estimates, it may be more like 23–46 cm, which would double the predicted rise (Meier, 2002). Even this may be a conservative estimate based on the new observations of accelerating mass loss of the Greenland ice sheet discussed in the previous chapter. Just a 1 meter rise in sea level would displace about 100 million of

FIGURE 11.7 The upper air photo shows the shoreline of South Bethany, Delaware. The squares and circles indicate houses. The white lines are the projected shoreline in the next 60 years with the circles indicating destroyed houses. This is probably underestimated because the sea level rise is about twice as much as in their study. At the bottom is a house in West Hampton, New York, that has almost been destroyed by beach erosion (1993). (Images from the Heinz Center report on coastal erosion, 2000)

FIGURE 11.8 Houses flooded by 2006 high tides exacerbated by rising sea levels in Funafuti, Tuvalu Islands. (Photo courtesy of Monise Laafai, TuvaluIslands.com)

today's coastal population. Thirty of the world's largest cities lie near coasts that would be vulnerable to a 1 meter sea level rise. Residents of low-lying river deltas, such as the Mississippi, Nile and Ganges, are particularly at risk of a rising sea level. For example, Figure 11.9 shows the effects of a 0.5 to 1 meter rise of sea level on the Nile delta. For a 0.5 meter increase, 3.8 million people would be displaced and 1,800 km^2 of cropland would be lost, and for a 1 meter increase, 6.1 million people would be displaced with the loss of 4,500 km^2 of cropland. In both cases, Alexandria, Port Said, and several smaller towns would be covered.

The above scenarios assume a steady rise in sea level due to the thermal expansion of seawater and the melting of glaciers over the next 100 years. The new IPCC 4th Assessment Report predicts a sea level rise of only 0.18 to 0.59 meter by the end of the century. That estimate is almost surely too low. Like the above scenarios, it is based on computer simulations of melting glaciers and ice sheets, and the thermal expansion of seawater. It does include projections of the sea level rise due to increased ice flow from Greenland and Antarctica, but at the rates observed for 1993–2003. It does not, and could not, include the collapse

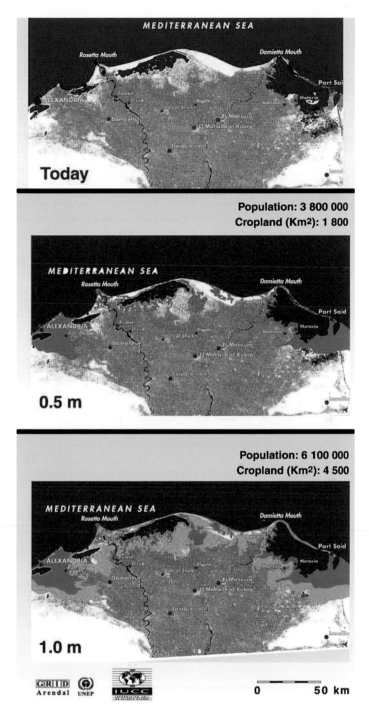

FIGURE 11.9 The effects of a 0.5-meter and a 1.0-meter rise in sea level on the Nile Delta. (Source: UN Environmental Program, GRID/Arendal Publ.)

of portions of the Antarctic or Greenland ice sheet from increased destabilization by melting as discussed in the last chapter. A new study finds that on timescales of present warming the rate of sea level change is proportional to the amount of warming above pre-Industrial Age temperatures, and that it is a good approximation for temperatures and sea level changes during the 20th century (Rahmstorf, 2007). When this is applied to future warming scenarios of the IPCC report, the projected sea level rise is 0.5 to 1.4 meters above the 1990 level by the year 2100.

One worrisome event occurred in the past. During the Bølling abrupt warming about 14,000 years ago (see Figures 5.9 and 5.10) sea level rose approximately 20 meters in about 400 years (Kienast et al., 2003), or an **average** rise of 5 cm/year (1 meter in 20 years). A 5 cm/year sea level rise requires about 15,000 km^3 of water delivered to the oceans each year. This implies a catastrophic break-up of at least part of an ice sheet during the last ice age. It is unlikely that the 20 meter rise was evenly distributed over the 400-year period. It is far more likely that the 20 meter rise occurred in spurts. There is no reason that a 1-meter sea level rise could not have occurred in a few years or less. If that happened to parts of the west Antarctic or Greenland ice sheets today we would be in real trouble. Figure 11.9 shows what just 0.5- and 1-meter sea level rise does to the Nile delta.

In fact, it is probably too late to prevent a substantial loss of glaciers and ice sheets in the future because of the current greenhouse gas content of the atmosphere, and the at least 1° C future warming due to the inertia of the climate system. If the West Antarctic ice sheet or its equivalent collapsed into the ocean it would raise sea level from 1 to 6 meters and completely inundate most of southern Florida (Figure 11.10), and large parts of other low-lying areas of the world. It would inundate, New York City, Washington, D.C., Calcutta, and most other ports in the world. Although it may be unlikely to happen catastrophically, it does illustrate what is in store for us in the future if melting continues to accelerate.

Figure 11.11a shows the global land area inundated and the population affected by sea level rises from 1 to 6 meters. Just a one-meter rise would affect 108 million people of today's population (Rowley, et al., 2007). It would inundate a land area over 1 million km^2, flooding most, if not all, of the world's ports and causing unimaginable economic disruption (Figure 11.11b). Although a 1-meter rise in a short period of time is unlikely today, it is possible by the end of the century.

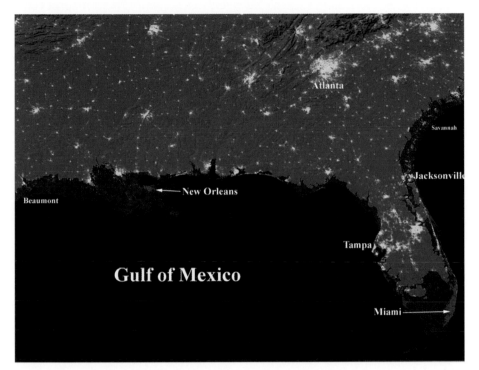

FIGURE 11.10 The effects of a 6-meter rise in sea level on the southeaster seaboard of the United States. The red and pink areas (high population) are under water. The white areas are major population centers. (Courtesy of Jeremy L. Weiss and Jonathan T. Overpeck, Dept of Geosciences, The University of Arizona, Tucson, AZ)

Chemistry

One the most important chemical changes that is occurring to the oceans is their acidification (Orr et al., 2005). About one-third of the 31 billion tons of CO_2 humans cause each year is being absorbed by the oceans. The additional amount of absorbed CO_2 from the increase in atmospheric CO_2 is reacting with the water (H_2O) to produce carbonic acid (HCO_3). Carbonic acid is corrosive to shells. The oceans are normally slightly alkaline, but now they are becoming acidic with the burden of absorbing more CO_2. The pH scale that measures acidity and alkalinity runs from 1 (the most acid) to 14 (the most alkaline). Today the ocean pH is 8.1 which is a 0.1 decrease since the industrial revolution began. Because the pH scale is logarithmic, the 0.1 decrease means that there is a 30% increase in hydrogen ions in the water. Depending on the rate of fossil fuel burning, the pH of ocean water near the surface is expected to drop to 7.7 or 7.9 by 2100. This is lower than at any time in at least the last 420,000 years (the extent of the glacial record where past alkalinity can be measured).

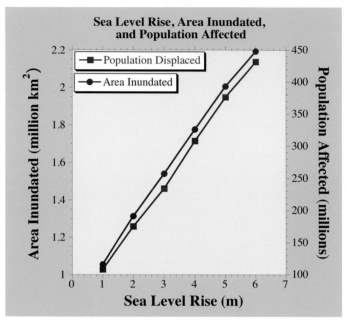

FIGURE 11.11a *This graph shows the area inundated and the global population affected by a 1 to 6 meters rise in sea level. A rise of only 1 meter affects about 108 million people and inundates about 1.05 million km². A 6-meter rise affects over 431 million people and inundates about 2.3 million km². (Data source: Rowley, et al., Risk of rising sea level to population and land area, EOS, 88, no. 9, 27 Feb. 2007)*

Changing the oceans from alkaline to acidic will probably have profound effects on sea life, particularly coral and calcareous-shelled organisms. It could destroy or greatly reduce plankton and disrupt the present food chain.

There has been a substantial reduction of dissolved oxygen (O_2) in the oceans above a depth of 3 km (Plattner et al., 2002; Keeling and Garcia, 2002). Dissolved O_2 is produced by biological processes, and it is removed in the subsurface waters by the chemical processes in the sinking organic matter. Changes in ocean circulation are probably the main cause of the decrease in dissolved oxygen. The change is a slowdown in the overturning circulation in a north–south direction due to warming of the oceans. The decrease in dissolved O_2 has led to a decrease in the ocean uptake of CO_2 by about 2.5 billion metric tons per year (Joos et al., 2003).

In the high North Atlantic the salinity of the water is decreasing because of the addition of fresh water from melting glaciers and increased rainfall, both due to global warming. Salinity measurements in the East Irminger and Labrador Seas, the Denmark Strait, and the Faroe–Shetland Channel show a continuous decrease over the past 35 years (Dickson et al., 2002).

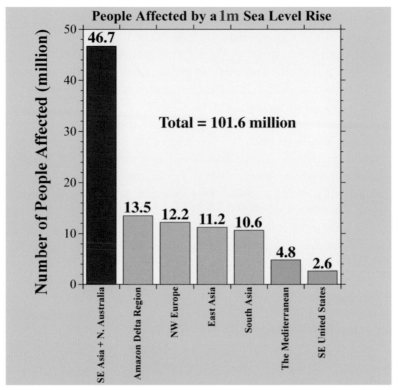

FIGURE 11.11b This bar chart shows the population affected by only a 1 meter rise in sea level. Southeast Asia plus Northern Australia (red) have over 45 million people affected. The orange bars have over 10 million people affected, and the green bars have over 2 million people affected. (Data source: Rowley, et al., Risk of rising sea level to population and land area, EOS, 88, no. 9, 27 Feb. 2007)

Since the 1960s the Arctic river discharge into the Arctic Ocean has increased at a rate of 8.73 km³/year (Wu et al., 2004). The flux of fresh water into the Bering Sea, which eventually reaches the Arctic Ocean, has increased even more to about 2,500 km³/year (Woodgate and Aagaard, 2005). About 20,000 km³ of fresh water has been added to the North Atlantic since 1970 (Curry and Mauritzen, 2005). The fresh water lowers the salinity, and eventually this diluted seawater reaches the North Atlantic to lower its salinity. It has been proposed that if the salinity gets too low, there is a possibility that the ocean current (called the subpolar gyre) carrying the warm water that keeps northern Europe temperate, could be disrupted.

In the past, drastic cooling in the North Atlantic region has occurred when the gyre shut down because of enormous influxes of fresh water from the catastrophic draining of glacial lakes (see Chapter 5). We do not know the threshold conditions that led to a complete shutdown of the subpolar gyre. We do know that previous shutdowns were the result of much larger fresh water influxes than are occurring today, but we do not know the minimum influx of fresh water required for a shutdown. In any event, a shutdown would not initiate a new ice age, as depicted in the movie "Day After Tomorrow."

New computer models of this mechanism indicate that a very rapid collapse seems unlikely under current greenhouse warming. Also, the thermohaline circulation is driven by wind, and as long as the wind blows, the essential parts of the thermohaline circulation will continue to flow. Vast amounts of ice also appear to be required to supply the meltwater necessary to shut down the thermohaline circulation at the end of the last ice age, and most of this ice has melted long ago. The most sophisticated and realistic model simulations of global warming fail to drive the thermohaline circulation anywhere near collapse (Wu et al., 2004; Gent, 2001; Gregory et al., 2005). In one case the projected warming only slowed the circulation 25 to 30% very gradually, and in the other it was 10 to 50%, also very gradually. This is a long way from total collapse. Eleven different computer runs included five "intermediate complexity" models that ran for 140 simulation years, quadrupling the concentration of greenhouse gases in the process (Gregory et al., 2005). In order to collapse the circulation, the flow of fresh water into the North Atlantic has to be 10 times faster than current greenhouse simulations. The final impact of any cooling effect will depend on whether it outweighs the global warming that is driving it. Over time any decrease in temperature caused by a slowing of the Atlantic current will be swamped by a more general warming of the atmosphere, according to climate modelers. The best information available today indicates that a shutdown of the sub-polar gyre is highly unlikely.

Biology

As you might expect, changes in the oceans' temperature and chemistry are beginning to have profound impacts on marine life. As mentioned earlier, changing the oceans from an alkaline to an acidic condition is beginning to kill the plankton population with adverse consequences for the rest of the marine food chain.

Measurements from the SeaWiFS satellite (1997 to present) and the Coastal Zone Color Scanner (1979–1986) indicate that the oceans' primary production of chlorophyll is declining (Gregg et al., 2003) and has declined about 6.3% since the early 1980s. About 70% of the global 10-year decline occurred in high latitudes. The declines are the result of increased sea-surface temperatures and decreased atmospheric iron deposition (a nutrient that promotes chlorophyll growth) to the oceans. This decline corresponds to a sea-surface temperature increase of $0.2\,°C$, and the higher temperature inhibits nutrients at lower depths in the oceans from reaching the surface to support the oceans' primary production of chlorophyll. This indicates that the oceans' photosynthetic uptake of carbon is decreasing as a result of global warming and eventually will lead to increased warming as the carbon uptake decreases (see Chapter 7).

Another effect of atmospheric iron deposition in the oceans is on the amount of CO_2 in the atmosphere. An increase has no great effect on the CO_2 concentration, but a decrease in iron deposition results in a strong increase in atmospheric CO_2 concentration (Parekh et al., 2006). Consequently, if a warmer world is less dusty there will be a positive feedback amplifying an increase in CO_2.

Larger marine animals are also being seriously affected. Coral reefs are extremely intolerant to small rises in temperature that destroy the ecosystem of the reef with the result that they could become lifeless, bleached exoskeletons devoid of tropical fish and other reef-dependent creatures. Most corals die at temperatures above 30 to $32\,°C$. The bleaching is the result of the expulsion of a symbiotic algae (*zooxanthellae*) that provides the reef with its color, carbon, and ability to deposit calcium carbonate (limestone). Once the coral is bleached it cannot receive adequate nutrients or oxygen to stay alive. Coral reef destruction is also resulting from human activity that includes over-fishing, coastal development, nutrient runoff from agriculture and sewage, and sedimentation from logging on streams that feed into coastal waters. It has been estimated that 58% of the world's reefs were threatened by this type of human activity in the late 1990s (Bryant et al., 1998). This human activity, combined with ocean warming (also human caused), does not bode well for coral reefs. Coral-bleaching events are becoming more frequent and extensive due to ocean warming (McNeil et al., 2004). For example, on Easter Island 90% of the reefs have lost 96% of their symbiotic algae (Wellington et al., 2001), which is usually enough to cause coral death. High levels of bleaching have affected an average of 85–90% of the colonies to a depth of 10 meters.

Northeastern Australia (Queensland and New South Wales) had the hottest summer on record in 2006. As a consequence ocean temperatures rose to dangerous levels for coral and up to 50% bleached at certain inshore locations in the southern Great Barrier Reef. Warmer ocean temperatures also result in decreased concentrations of phytoplankton sending ripples through the food chain. The worst coral-bleaching event

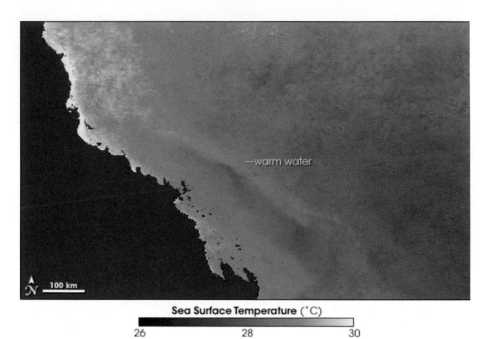

—warm water

Sea Surface Temperature (°C)

26   28   30

FIGURE 11.12 This image shows the sea-surface temperature distribution in the vicinity of the Great Barrier Reef of northeastern Australia between February 6 and 20, 2006. Pink and yellow tones show regions of warm shallow waters over the reef near the coast where coral bleaching was the most severe. (Courtesy of NASA from the Aqua satellite. Image by Norman Kuring and Robert Simmon, based on data provided by Scarla Weeks, University of Queensland)

occurred in 2002 and phytoplankton decrease resulted in 50% of seabird chicks on Heron Island in the southern Great Barrier Reef starving to death because the adult birds were unable to find enough fish. Figure 11.12 illustrates the sea-surface temperatures between February 6 and 20, 2006.

The warming oceans are also causing fish populations (at least in the North Sea) to migrate to higher latitudes or deeper (cooler) waters (Perry et al., 2005). Data compiled on 36 of the most common bottom-dwelling fish in the North Sea indicates that two-thirds of the populations moved toward cooler water, going north or to deeper waters, or both, over the past 25 years. This included both commercial and non-commercial species. Because species are redistributing at different rates, the shifts could disrupt ties within ecosystems. If the migrating species suddenly encounter new predators with which they are unfamiliar, the balance of the ecosystem could change with adverse effects on the species. Already heavy commercial fishing has pushed some species in the North Sea to the brink of extinction. Fish migration due to a warming ocean could hamper the revival of over-fished species, disrupt ecosystems and lead to extinctions of many species. Already species that shifted their distributions have faster life cycles and smaller sizes than non-shifted species, and continued global warming will only exacerbate the problem.

The migration of species due to global warming is not confined to the oceans (Root et al., 2003; Parmesan and Yohe, 2002). In studies of over 1,700 species of animals and plants around the world, 81% showed migrations to higher latitudes consistent with those predicted by global warming. The range shifts averaged about 6 km per decade toward the poles. Furthermore, they found that the advancement of spring events (budding of vegetation, for example) had increased by over 2 days per decade. Several bird species of both short- and long-distance migrations have advanced their spring migrations in response to a warming climate (Jonzen et al., 2006). Climate change is also affecting the survival of species. It is estimated that, for a sample of over 1,100 terrestrial species from Mexico to Australia, the extinction probability for different amounts of increased warming is 18% for 0.8–1.7 °C, 24% for 1.8–2.0 °C and 35% above 2.0 °C (Thomas et al., 2004). If this is typical then we could be heading for another major extinction event just due to global warming.

Remarkably, the recent rapid temperature rise is beginning to drive evolution, at least in some species (Bradshaw and Holzapfel, 2006). Some species of small mammals, plants, and birds have adapted to altered seasonal events, rather than direct effect of temperature, producing genetic changes. These genetic changes have affected the timing of major events such as when to develop, when to reproduce, when to enter dormancy, and when to migrate. This adaptation may help some species to survive climate change; in fact, it may have been responsible for the survival of at least some species during past extinction events.

Local Climate

The increased North Atlantic Ocean temperatures appear to be effecting the North American and European summer climate (Sutton and Hodson, 2005). There is evidence that the frequency of droughts (McCabe et al., 2004) and the frequency of heat waves (Cassou et al., 2005) are both sensitive to the sea-surface temperature of the North Atlantic. The observed increase in water temperature will probably increase the frequency and intensity of droughts and heat waves. The severe European heat wave of 2003 may be the direct or indirect result of this increased water temperature.

EXTREME WEATHER EVENTS

The warming oceans also have an affect on extreme weather events like, hurricanes, torrential rainfall and flooding, droughts and desertification, and even heat waves. The frequency and/or intensity of extreme weather conditions is increasing and in some cases is unprecedented.

The increase in greenhouse gases and global warming have apparently also resulted in a major increase in precipitation during the last 150 years (Treydte et al., 2006). A reconstruction of precipitation in central Asia from tree-ring data shows a dramatic increase over the past century compared to the previous thousand years. This agrees with other observations around the world, and with computer models. Also, recent changes in precipitation patterns exceed the range of natural variability for the past several hundred to a thousand years. In fact, the global precipitation increased 1% during the 20th century. In addition, their study suggests that not only has the precipitation increased, but that observed drying trends in the subtropics, and wetting trends in the tropics and high latitudes agree with model predictions. Regional droughts have also increased.

Global warming is also effecting the Earth's atmospheric circulation. An enormous wind circulation pattern over the Pacific Ocean called the Walker Circulation has begun to weaken (Vecchi et al., 2006). It started to weaken in the mid–1800s and has continued to weaken since then. The observed slowdown is consistent with climate simulations that use increases in greenhouse gases, but fails when only natural influences are used. This indicates that human-caused greenhouse gas emissions are responsible. The weakening circulation could alter the climate and the marine food chain in the region.

Flooding and Water Runoff

As the temperatures rise there is more evaporation from the warming oceans, and, therefore, one could expect heavier precipitation in the form of rain and snow storms. Consequently, the frequency of major floods might also increase. The data for the last 20 years seems to bear that out; the number of major floods has increased between 1986 and 2000, but the increase has been particularly dramatic since 2000. Figure 11.13 shows the increase in major floods in 5-year intervals from 1986 to 2005 (the most reliable data) together with the average temperature anomaly over the same 5-year intervals.[3]

---

[3] A major flood is one that results in significant to extreme damage.

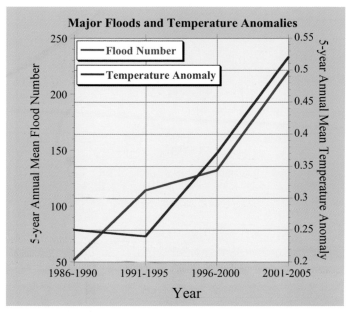

FIGURE 11.13 This graph shows the correlation of rising temperatures and increasing incidents of the major floods throughout the world for 5-year intervals from 1986 to 2005. (Data courtesy the Dartmouth Flood Observatory)

Water runoff is a measure of sustainable water availability; the larger the runoff, the more available the water for drinking and irrigation. Computer studies have shown that there will likely be major changes in water runoff due to global warming by 2050 (Milly et al., 2005), which suggests that there will be more flooding in regions of major runoff and draught-like conditions in regions of reduced runoff. The simulations suggest 10–40% increases in runoff in eastern equatorial Africa, the La Plata basin in eastern South America, and high latitudes of North America and Eurasia. About 10–30% decreases in runoff will occur in southern Africa, southern Europe, the Middle East, and mid-latitude western North America. These changes in sustainable water availability will have significant regional-scale impacts on their economies and ecosystems.

Hurricanes

Another deadly weather system has increased: the strength of hurricanes. They derive their strength in large part from the ocean's heat. We have already seen that ocean temperatures are increasing rapidly, so one might expect that the frequency of strong hurricanes would also increase—and that appears to be exactly what is happening (Webster et al., 2005; Emanuel, 2005). The tropical ocean sea-surface temperatures increased

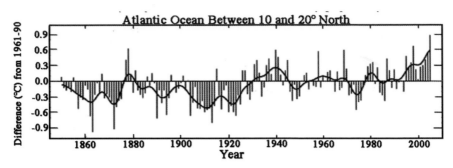

FIGURE 11.14 *The variation sea surface temperature in the Atlantic Ocean from 10 to 20 north latitude, where most hurricanes form, is shown in this graph. The sea surface temperature has risen significantly since about 1995. (From UN IPCC 4th Assessment 2007)*

about 0.5 °C between 1970 and 2004. Figure 11.14 shows the increase in the ocean surface temperature between 10 and 20 degrees north latitude where hurricanes form. There has been a marked increase from about 1995 to 2005. The particularly high value in 2005 coincides with the extreme hurricane frequency and strength that year.

In the North Atlantic and eastern Pacific tropical storms are called hurricanes, but in the eastern Pacific they are called typhoons, and in the Indian Ocean they are called cyclones. In order to determine if the frequency or strength of hurricanes has increased or not, it is necessary to study all of these storms wherever they occur in the world. The strength of hurricanes is rated from 1 to 5 on the Saffir–Simpson scale. Their speeds range in meters/sec, respectively, from 33 to 43, 43 to 50, 50 to 56, 56 to 67, and greater than 67. A worldwide study of hurricanes from 1970 to 2004 shows that the overall frequency and absolute strength of hurricanes have not changed very much over the 35-year length of the study (Webster et al., 2005). However, the frequency of the most destructive Category 4 and 5 hurricanes has increased significantly. Figure 11.15 shows the number and percentage of Category 4 and 5 hurricanes for the 1975–1989 and 1990–2004 periods in various parts of the world. The average number of Category 4 and 5 hurricanes increased from about 28 to 45, or an average 13% increase between the two periods. Also in 2005, for the first time on record, a hurricane occurred in the South Atlantic and slammed into Brazil causing 16 deaths (Pezza and Simmonds, 2005).

The overall frequency of all categories of hurricanes has increased in the North Atlantic, with 2005 having a record 27 named tropical storms, with 13 of them hurricanes. One tropical storm occurred one month after the "official" end of the hurricane season on November 30. Some meteorologists (e.g. Goldenberg et al., 2001) attribute this increase in frequency to a weather and ocean condition called the Atlantic Multi-

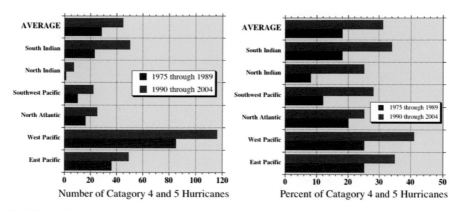

FIGURE 11.15 These bars charts show the increasing number (left) and percentage (right) of Category 4 and 5 hurricanes for the periods 1975 through 1989 and 1990 through 2004 in various tropical oceans. New studies indicate that the average frequency of hurricanes will also increase. (Data from Table 1 in Webster et al., 2005)

decadal Oscillation (AMO). New studies contradict the claim that natural cycles are responsible for the increase in Atlantic hurricane activity since 1995 (Trenberth and Shea, 2006; Mann and Emanuel, 2006). They show that global warming accounted for about half of the extra warmth in the waters of the tropical North Atlantic in 2005, but natural cycles such as the Atlantic Multi-decadal Oscillation were only a minor factor. The results show that the AMO is actually much weaker now than it was in the 1950s, when Atlantic hurricanes were also quite active. This indicates that, **on average**, hurricane seasons will become more active as global temperatures rise. However, it does not mean that every year will see a higher frequency and strength of hurricanes, because they also depend on favorable upper-level winds, the presence or absence of El Niños, and other factors.

The 2006 Atlantic hurricane season produced fewer hurricanes than predicted by scientists. There were only 5 hurricanes (about average), two of which were category 3, and none made landfall. The reason is that an El Niño developed in the Pacific and produced intense upper level wind shear over the Atlantic. This wind shear disrupted tropical storms preventing them from becoming hurricanes. Furthermore, the increased atmospheric loading of Saharan dust over the North Atlantic during the hurricane season producing a cooling of the north Atlantic Ocean suppressing hurricane formation (Lau and Kim, 2007).

Droughts and Desertification

Droughts and desertification are becoming serious problems in certain parts of the world. Satellite data acquired between 1979 and 2005 show that the subtropical atmosphere in both hemispheres has warmed enough to cause it to expand at altitudes occupied by the jet streams (Fu et al., 2006). As a result, it has pushed both jet streams about 110 km closer to the poles causing areas already stressed by drought to get even drier. This is affecting areas that include the American southwest, southern Australia, and the Mediterranean basin. It is also moving the Sahara Desert farther north.

Mongolia's grassland plains are becoming a desert. Although bad agricultural management is partly responsible, the changing climate is mostly to blame in part due to the melting glaciers on the Tibetan Plateau. Mongolia's rainfall has declined significantly as the temperatures have risen at twice the global average. About 2,000 km^2 of grassland are turning into desert each year. West of inner Mongolia in the Gansu province the river systems have dried up. Now the Yellow River fails to reach the sea more than half the year on average. Sand is burying villages, irrigation wells, and roads. There were only eight dust storms in the 1960s, but in the 1980s there were 14, and in the 1990s there were 23. In 2000 alone seven strong dust storms swept through Beijing, and from January to May of 2006 there were 13. One dust storm in April 2006 swept across an eighth of China and reached Korea and Japan. It dumped a record 336,000 tons of dust on Beijing, resulting in extreme and dangerous air pollution.

Spain and Portugal (2005) are suffering their worst droughts since records began, and in western France water levels are at their lowest since the major drought of 1976. Drought conditions now stretch from North Africa to the French capital, causing billions of euros worth of damage as crops die, rivers dry up, and pastures turn to dust.

Computer models indicate that, in the future, dune fields of southern Africa will be reactivated (the sand will become exposed and move) due to global warming (Thomas et al., 2005). The authors state, "the general trend and the magnitude of possible changes in the erodibility and erosivity of dune systems suggests that the environmental and social consequences of these changes will be drastic."

Other severe droughts are occurring in Africa, South America, the western part of the United States, and Australia. In the western United States, rainfall and snow packs are decreasing causing the onset of water shortages. The average snow pack in the Cascade Mountains has declined 50% since 1950, and will decrease another 50% in about 30 years if we do not address the problems of climate change now. That snow provides the drinking water and powers the hydroelectric dams that provide energy (Service, 2004; Mote, 2003). Parts of the southwestern United States are

2080-2099

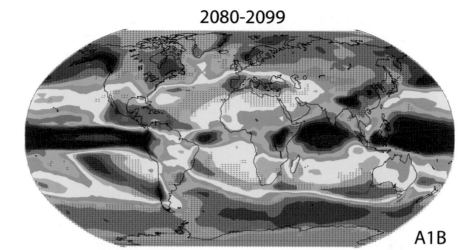

A1B

-0.5 -0.4 -0.3 -0.2 -0.1 0 0.1 0.2 0.3 0.4 0.5

Annual Mean Precipitation Change (mm/day)

FIGURE 11.16 Computer simulation of the future annual mean precipitation from 2080 to 2090. The yellow and red areas are regions of decreased precipitation that are drought-prone, and the light and dark blue regions of excess precipitation are prone to flooding or heavy snow. See text for discussion. (From UN IPCC 4thAssessment 2007)

in severe to extreme drought conditions, greatly exacerbating fire danger; and wildfires in the American west and Canada have increased due to global warming. The pattern of warming of the North Pacific surface waters correlates with the intense Northwest wildfires (Liu, 2006). A recent study (Westerling et al., 2006) shows that warmer temperatures appear to be increasing the duration and intensity of the wildfire season in the western United States. Compared to the period from 1970 to 1986, longer and warmer summers have caused a four-fold increase of major wildfires. Also during the same period there has been a six-fold increase in the area burned.

Figure 11.16 is a computer simulation of the future annual mean precipitation from 2080 to 2090. The yellow and red areas are regions of drought, and the light and dark blue regions of excess precipitation prone to flooding or heavy snow. In general, the areas of drought predicted for the end of the century, such as the Amazon, American southwest, southern Europe, and Australia, are happening today. Also the areas predicted for more rain and flooding are generally happening today. This suggests that the weather effects of global warming are accelerating faster than predicted by computer simulations.

Heat Waves

Extreme heat waves and their durations have also increased. Record-setting heat waves have recently occurred in India, China, and other parts of the world. A notable record-setting heat wave occurred in Europe in the summer of 2003. Although this particular heat wave cannot be unambiguously attributed to global warming, it may portend European summers in the future. During August of 2003 the temperature in southern England reached a high of 38 °C and in parts of France and Switzerland it reached 41 °C. At the Cordoba airport in Spain the temperature reached 45 °C (114 °F). The English temperature was the highest ever since records have been kept. The heat wave lasted over two weeks with both day-time and night-time temperatures that were abnormally high. As a result, over 30,000 people died of heat-related conditions, and 80% were over the age of 75. In France alone about 15,000 people died. Multiproxy reconstructions of monthly and seasonal

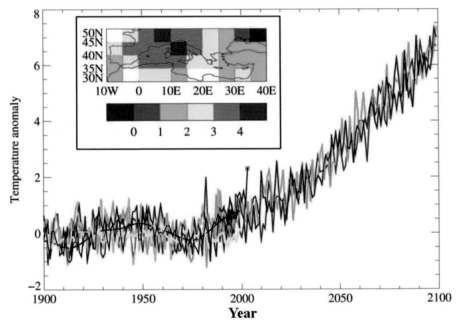

FIGURE 11.17 *This graph shows the European June–August temperature anomalies relative to the 1961–1990 mean for the region shown in the insert, which also shows the observed summer 2003 temperature anomalies. The thin black line is the observed temperatures and the thick black line is the filtered (average) temperature. The European heat wave of 2003 is shown with a star. The colored lines are four computer simulations including human and natural forcings and the yellow line is only natural forcings. The colored lines beyond about 2005 are computer extrapolations to 2100 for a doubling of CO_2 over the pre-industrial level. (From Brown et al., 2005; Fig. 1)*

temperatures for Europe back to 1,500 show that in the late 20th and early 21st centuries the climate was warmer (at >95% confidence level) than at any time during the past 500 years, and that 2003 was by far the hottest summer ever recorded (Luterhacher et al., 2004). This strongly suggests that the heat waves are related to global warming. Furthermore, computer models suggest that in future decades summers will be an average of 4 °C hotter (Beniston, 2004). Also the frequency, intensity, and duration of extreme heat waves in both Europe and North America will increase in the 21st century (Meehl and Tebaldi, 2004). Figure 11.17 shows temperature observations and computer models that suggest that the 2003 European heat wave will be the summer norm in about 40 years.

SUMMARY

Global warming is having large adverse impacts on the world's oceans: their level, chemistry, temperature, and life. These changes in turn are beginning to have dramatic environmental effects that are starting to have serious consequences for humans, including increased flooding and droughts, increases in the frequency of strong hurricanes, and increases in the frequency, duration and intensity of heat waves. There is no doubt that these problems will only increase in the future. If something is not done to combat these changes there will be "hell to pay." The world's oceans are the largest and most important Earth system. If it is damaged, then the rest of the world will also be damaged in ways we cannot fully predict.

12

WHAT'S IN STORE FOR US?

THE future consequences of global warming are the least known aspect of the problem. They are based on highly complex computer models that rely on inputs that are sometimes not well known or factors that may be completely unforeseen. Most models assume certain scenarios concerning the rise in greenhouse gases. Some assume that we continue to release them at the current rate of increase while others assume that we curtail greenhouse gas release to one degree or another. Furthermore, we are in completely unknown territory. The current greenhouse gas content of the atmosphere has not been as high in at least the past 650,000 years, and the rise in temperature has not been as rapid since civilization began some 10,000 years ago. What lies ahead for us is not completely understood, but it certainly will not be good, and it could be catastrophic.

We know that relatively minor climatic events have had strong adverse effects on humanity, and some of these were mentioned in previous chapters. A recent example is the strong El Niño event of 1997–1998 that caused weather damage around the world totaling $100 billion: major flooding events in China, massive fires in Borneo and the Amazon jungle, and extreme drought in Mexico and Central America. That event was nothing compared to what lies in store for us in the future if we do nothing to curb global warming. We currently face the greatest threat to humanity since civilization began.

WHAT IS THE CRITICAL TEMPERATURE FOR HUMANS?

This is the crucial, central question, but it is very difficult to answer (Mastrandea and Schneider, 2004). An even more important question is: "At what temperature and environmental conditions is a threshold crossed that leads to an abrupt and catastrophic climate change?" It is not possible to answer that question now, but we must be aware that in our ignorance it could happen in the not too distant future.

At least the question of a critical temperature is possible to estimate from studies in the current science literature. This has been done by the Potsdam Institute for Climate Impact Research, Germany's leading climate change research institute (Hare, 2005). According to this study, global warming impacts multiply and accelerate rapidly as the average global temperature rises. We are certainly beginning to see that now.

According to the study, as the average global temperature anomaly rises to 1 °C within the next 25 years (it is already 0.6 °C in the Northern Hemisphere), some specialized ecosystems become very stressed, and in some developing countries food production will begin a serious decline, water shortage problems will worsen, and there will be net losses in the gross domestic product (GDP). At least one study finds that because of the time lags between changes in radiative forcing we are in for a 1 °C increase before equilibrating even if the radiative forcing is fixed at today's level (Wetherald et al., 2001).

It is apparently when the temperature anomaly reaches 2 °C that serious effects will start to come rapidly and with brute force (International Climate Change Taskforce, 2005). At the current rate of increase this is expected to happen sometime in the middle of this century. At that point there is nothing to do but try to adapt to the changes. Besides the loss of animal and plant species and the rapid exacerbation of our present problems, there are likely to be large numbers of hungry, diseased and starving people, and at least 1.5 billion people facing severe water shortages. GDP losses will be significant and the spread of diseases will be widespread (see below). We are only about 30 years away from the 440 ppm CO_2 level where the eventual 2 °C global average temperature is probable.

When the temperature reaches 3 °C above today's level, the effects appear to become absolutely critical. At the current rate of greenhouse gas emission, that point is expected to be reached in the second half of the century. For example, it is expected that the Amazon rainforest will become irreversibly damaged leading to its collapse, and that the complete destruction of coral reefs will be widespread. As these things are already happening, this picture may be optimistic. As for humans, there will be widespread hunger and starvation with up to 5.5 billion people living in regions with large crop losses and another 3 billion people with serious water shortages. If the Amazon rainforest collapses due to severe drought it would result in decreased uptake of CO_2 from the soil and vegetation of about 270 billion tons, resulting in an enormous increase in the atmospheric level of CO_2. This, of course, would lead to even hotter temperatures with catastrophic results for civilization.

A Regional Climate Change Index has been established that estimates the impact of global warming on various regions of the world (Giorgi, 2006). The index is based on four variables that include changes in surface temperature and precipitation in 2080–2099 compared to the period 1960–1979. All regions of the world are affected significantly, but some regions are much more vulnerable than others. The biggest impacts occur in the Mediterranean and northeastern European regions, followed by high-latitude Northern Hemisphere regions and Central America. Central America is the most affected tropical region followed by southern equatorial Africa and southeast Asia. Other prominent mid-latitude regions very vulnerable to global warming are eastern North America and central Asia.

It is entirely obvious that we must start curtailing greenhouse gas emissions now, not 5 or 10 or 20 years from now. Keeping the global average temperature anomaly under 2 °C will not be easy according to a recent report (Scientific Expert Group Report on Climate Change, 2007). It will require a rapid worldwide reduction in methane, and global CO_2 emissions must level off to a concentration not much greater than the present amount by about 2020. Emissions would then have to decline to about a third of that level by 2100. Delaying action will only insure a grim future for our children and grandchildren. If the current generation does not drastically reduce its greenhouse gas emission, then, unfortunately, our grandchildren will get what *we* deserve.

213

THREE CRITICAL CONSEQUENCES

There are three consequences that have not been discussed in previous chapters but could have devastating impacts on humans: food production, health, and the economy. In a sense, all of these topics are interrelated, because they affect each other.

Food Production

Agriculture is critical to the survival of civilization. Crops feed not only us but also the domestic animals we use for food. Any disruption in food production means a disruption of the economy, government, and health. The increase in CO_2 will result in some growth of crops, and rising temperatures will open new areas to crop production at higher latitudes and over longer growing seasons; however, the overall result will be decreased crop production in most parts of the world.

A 1993 study of the effects of a doubling of CO_2 (550 ppm) above pre-industrial levels shows that there will be substantial decreases in the world food supply (Rosenzweig et al., 1993). In their research they studied the effects of global warming on four crops (wheat, rice, protein feed, and coarse grain) using four scenarios involving various adaptations of crops to temperature change and CO_2 abundance. They found that the amount of world food reduction ranged from 1 to 27%. However, the optimistic value of 1% is almost certainly much too low, because it assumed that the amount of degradation would be offset by more growth from "CO_2 fertilization." We now know that this is not the case, as explained below and in Chapter 7. The most probable value is a worldwide food reduction between 16 and 27%. These scenarios are based on temperature and CO_2 rises that may be too low, as discussed in Chapter 7. However, even a decrease in world food production of 16% would lead to large-scale starvation in many regions of the world.

Large-scale experiments called Free-Air Concentration Enrichment have shown that the effects of higher CO_2 levels on crop growth is about 50% less than experiments in enclosure studies (Long et al., 2006). This shows that the projections that conclude that rising CO_2 will fully offset the losses due to higher temperatures are wrong. The downside of climate change will far outweigh the benefits of increased CO_2 and longer growing seasons. One researcher (Prof. Long) from the University of Illinois put it this way:

> Growing crops much closer to real conditions has shown that increased levels of carbon dioxide in the atmosphere will have roughly half the beneficial

effects previously hoped for in the event of climate change. In addition, ground-level ozone, which is also predicted to rise but has not been extensively studied before, has been shown to result in a loss of photosynthesis and 20 per cent reduction in crop yield. Both these results show that we need to seriously re-examine our predictions for future global food production, as they are likely to be far lower than previously estimated.

Also, studies in Britain and Denmark show that only a few days of hot temperatures can severely reduce the yield of major food crops such as wheat, soy beans, rice, and groundnuts if they coincide with the flowering of these crops. This suggests that there are certain thresholds above which crops become very vulnerable to climate change.

The European heat wave in the summer of 2003 provided a large-scale experiment on the behavior of crops to increased temperatures. Scientists from several European research institutes and universities found that the growth of plants during the heat wave was reduced by nearly a third (Ciais et al., 2005). In Italy, the growth of corn dropped by about 36% while oak and pine had a growth reduction of 30%. In the affected areas of the mid-west and California the summer heat wave of 2006 resulted in a 35% loss of crops, and in California a 15% decline in dairy production due to the heat-caused death of dairy cattle.

It has been projected that a 2 °C rise in local temperature will result in a $92 million loss to agriculture in the Yakima Valley of Washington due to the reduction of the snow pack. A 4 °C increase will result in a loss of about $163 million.

For the first time, the world's grain harvests have fallen below the consumption level for the past four years according to the Earth Policy Institute (Brown, 2003). Furthermore, the shortfall in grain production increased each year, from 16 million tons in 2000 to 93 million tons in 2003.

These studies were done in industrialized nations where agricultural practices are the best in the world. In developing nations the impact will be much more severe. It is here that the impact of global warming on crops and domestic animals will be most felt. In general, the world's most crucial staple food crops could fall by as much as one-third because of resistance to flowering and setting of seeds due to rising temperatures. Crop ecologists believe that many crops grown in the tropics are near, or at, their thermal limits. Already research in the Philippines has linked higher night-time temperatures to a reduction in rice yield. It is estimated that for rice, wheat, and corn, the grain yields are likely to decline by 10% for every local 1 °C increase in temperature. With a decreasing availability of food, malnutrition will become more frequent accompanied by damage to the immune system. This will result in a greater susceptibility to

215

spreading diseases. For an extreme rise in global temperature ($\geqslant 6\,°C$), it is likely that worldwide crop failures will lead to mass starvation, and political and economic chaos with all their ramifications for civilization.

Health

Rising temperatures will result in the spread of disease (Patz et al., 2005). The incidence of certain diseases depends to a large extent on the climate. Diseases that are now found in the tropics will spread to higher latitudes and greater altitudes as the climate warms. Those that occur in subtropical and temperate regions for only short periods each year will afflict residents for longer durations as warming intensifies.

There are a number of tropical diseases that are likely to spread northward as the climate warms (McMichael et al., 2003; Martens et al., 1995). These include malaria, dengue fever, schistomiasis, onchoncerciasis, lymphatic filariasis, sleeping sickness, leishmaniasis, chagas disease, and yellow fever. Currently, these diseases infect a total of about 800 million people, but the disease with the greatest potential for dissemination to higher latitudes is malaria. Figure 12.1 shows the potential risk of malaria epidemics for an increase of the global mean temperature of only $1.2\,°C$ compared for the risk during the 1931–1981 baseline climate. According to this projection, much of North America and Europe are at risk of large outbreaks of the disease with only moderate amounts of global warming.

As the climate warms, human populations will become far more vulnerable to heat-related mortality, air pollution-related illnesses, infectious diseases, and malnutrition. Areas of increased rainfall will become much more susceptible to the spread of waterborne and foodborne disease. Increased local rainfall will also make it easier for the insects and animals that carry some human diseases to flourish. At present about 9 million cases of waterborne disease occur each year in the United States where most people have access to treated water. Global warming will almost certainly increase that number. The World Health Organization estimates that currently 150,000 people die annually from the climate changes that have taken place in the past 30 years, and projects that millions of people will die from climate-rated diseases in the coming decades.

In fact, the spread of disease has already begun. Malaria has quadrupled between 1995 and 2000 due, at least in part, to warmer climates. Malaria is reappearing both north and south of the tropics. It is showing up more frequently in the United States, and has returned to the Korean peninsula, parts of southern Europe, Russia, and to the coast of South Africa along the Indian Ocean.

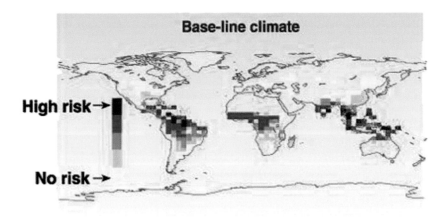

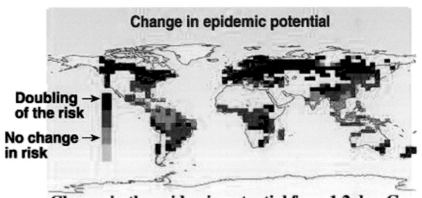

Change in the epidemic potential for a 1.2 deg. C increase in the global mean temperature above the 1931-1980 baseline climate.

FIGURE 12.1 These maps show the risk for the spread of Malaria with only a 1.2 °C increase in the global mean temperature above today's value. (Modified from Martens et al., 1995)

The spread of disease is not confined to humans. There has been an explosion of bark beetle infestation in the western United States and Canada. The rapid spread of the bark beetle is the result of warmer winters. This disease has killed millions of square kilometers of forests and made the western forests a tinderbox. The massive forest fires in Arizona and southern California during the summer of 2003 were in large part due to dry, dead trees caused by bark beetle infestation.

The combination of spreading diseases and decreases in crop production, together with all the other adverse manifestations of a warming Earth, will surely put an enormous strain on governments and the world economy. The largest effects will be felt by developing countries where we can expect many failed governments. All of this can be expected with only a 2 °C rise in the global mean temperature.

A recent study of Harlequin frogs in tropical America illustrates how global warming can cause the spread of disease and wreak havoc among a species (Pounds et al., 2006). About two-thirds of their species have died out because of a pathogenic fungus. Outbreaks of the fungus in mountainous regions of tropical America have occurred because night-time temperatures in these regions have shifted to more optimum conditions for the fungus. At the same time increased cloudiness has prevented the frogs from finding "thermal refuges" from the pathogen and, as a result of these climate changes, the frogs are nearing extinction. This example shows how the spread of disease due to global warming can seriously disrupt a population, including our own.

Economics

There is a steep price to pay for the consequences of global warming. One way to access the cost of global warming is to examine the insurance costs due to natural disasters. The insurance industry has yearly revenues of $3 trillion, making it three times the size of the oil industry. They are very good at analyzing risks and assessing costs. The numbers they are getting because of climate change are not looking good for their business. During the past century the frequency and impact of weather disasters have been steadily climbing, but there has been a dramatic increase in the last 50 years that coincides with the rise in greenhouse gases and temperatures. In the 1950s there were less than 200 weather-related disasters. However, that figure had jumped to over 1,600 in the 1990s. During the same period the economic losses caused by weather events increased 10 times, from $4 billion in the 1950s to $40 billion in the 1990s. As a result, the insurance industry's responsibility for covering weather-related losses has risen enormously. In the 1950s, insurance losses due to weather disasters were negligible, but by the 1990s they had increased to $9.2 billion a year. By the end of 2002, losses as a result of natural disasters appeared to be doubling every decade. They reached one trillion dollars in the past 15 years and will continue to rise as global temperatures rise (UN Environment Programme's Finance Initiatives, 2003). The year 2004 was the costliest year on record for the global insurance industry because of natural disasters, with losses of about $52 billion. Global reinsurers, the companies that accept the ultimate risk by covering insurers, paid out claims about 250% higher than the previous record.

Munich Re, one of the world's biggest reinsurers, had economic losses for 2004 of about $168 billion from natural disasters before the great tsunami at the end of that year. This is nearly double the average of the past 10 years and they expect damage from natural disasters to rise "exponentially" in the coming years because of global warming. Also that year, Lloyd's of London reported preliminary damage estimates of $13 billion for Asian floods. The year's total of about 650 natural disasters was about the same as 2003, but the damage was on a larger scale due to more severe flooding and other damage at least in part due to global warming. The Association of British Insurers warned that the cost of worldwide storms is likely to increase by two-thirds to about $27 billion a year in future decades.

Global warming is beginning to influence financial markets. According to Joachim Faber, the CEO of Allianz Global Investors, "In the interest of our clients and shareholders, we are obligated to take these risks into account when making decisions on insurance underwriting, investments, or credit." In an effort to price global warming risks into financial markets, Allianz has urged financial managers and analysts to evaluate their client portfolios for risks associated with global warming. It looks like insurance costs in areas at risk from natural events triggered by global warming will be going up, and that some areas at great risk will not even be insured.

A new 700-page report by Sir Nicholas Stern (Stern, 2006), chief of the British economic service and a former chief economist of the World Bank, evaluated scientific estimates of the consequences of global warming from an economic perspective. The basic finding was that, unless checked, global warming will have a devastating effect on the world economy. The costs related to global warming could consume as much as 20% of the world's gross domestic product (GDP). These economic costs would be on a scale similar to the economic Great Depression and World Wars I and II during the early part of the 20th century. The big difference is that in the 1930s the world population was only about 2 billion compared with about 8 billion when these global warming consequences start occurring. The estimated cost of cutting greenhouse gases to mitigate the problem is about 1% of the global GDP, a small price to pay. In other words, it would cost only 1% of the world GDP to be 20% richer than we otherwise would be. In fact, if you take the present value of the benefits over the coming years of taking action to stabilize greenhouse gases by 2050, and then deduct the costs, you actually gain a "profit" of about $2.5 trillion.

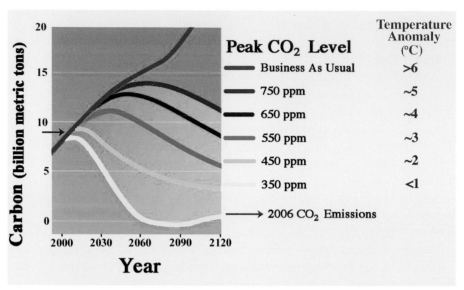

FIGURE 12.2 *This diagram shows the peak CO_2 levels for various emission scenarios and the approximate global average temperature anomalies that result from these levels. The red arrow indicates the 2007 level of 383 ppm. Therefore, the level to keep the temperature anomaly under 1 °C has already been exceeded and we are fast approaching the critical level of about 450 ppm. (Modified from John P. Holdren, Director, The Woods Hole Research Center, Teresa & John Heinz Professor of Environmental Policy, Harvard University and President, American Association for the Advancement of Science)*

GLOBAL WARMING SCENARIOS

The future of global warming and the future of humanity depend on limiting the peak value of greenhouse gases. That, in turn, depends on how soon and how strongly we curb our greenhouse gas emissions. Figure 12.2 is a schematic diagram showing the peak CO_2 levels for various emission scenarios together with the approximate global average temperature anomalies that result from these levels. The present level is 383 ppm (red arrow) and increasing at a rate of 2 ppm per year. We have already passed the level (~ 350 ppm) where we could keep the temperature anomaly under 1 °C, and we are fast approaching the critical level of about 450 ppm where a 2 °C anomaly is inevitable. Keep in mind that these temperatures could be on the low side. The other peak CO_2 levels would certainly have extremely adverse effects on humanity.

The future consequences of global warming are almost impossible to predict because they depend on our actions and unknown environmental conditions. However, there appears to be three general scenarios that are possible. One of these is our best possible outcome. Another would cause at least serious disruptions and could prove calamitous to humanity. The other one would almost surely be catastrophic to civilization. The three scenarios are: (1) stabilize atmospheric greenhouse gases by cutting total emissions by about 60%; (2) continue emitting greenhouse gases at the present rate with a 2 to 4.5 °C increase in the global average temperature; and (3) trigger an abrupt climate change that propels us into a "hot house".

Stabilize Atmospheric CO_2

The best possible climate change scenario is a 60% reduction of greenhouse gas emissions by the world community. This would eventually stabilize the greenhouse gas content of the atmosphere at some level that depends on how soon the reduction occurs. Even if all human greenhouse gas emissions went to zero today, the average global temperature could still rise about another 1 °C by mid-century (Wetherald et al., 2001). It is estimated that if the CO_2 content reaches 440 ppm, a global average temperature rise of 2 °C is probable. Today the CO_2 content is 383 ppm and rising at an average rate of 2 ppm per year. Therefore, we must make drastic cuts in our CO_2 emissions within the next 30 years so that we can slow the rise in the atmospheric CO_2 content. Failing that we should be planning strategies to deal with the consequences that a 2 °C rise will bring. This is why it is so important to start large reductions in greenhouse gas emissions now. We could still be in for a large sea level rise if the Greenland and west Antarctic ice sheets begin collapsing at higher rates.

Business As Usual

If we keep emitting greenhouse gases at or near the current rate then we are in for at least a 3.5 °C rise, and probably as high as 6 °C rise, in global average temperature by the end of this century. The higher value is less likely, but if a 6 °C rise did happen within 100 years it would be the equivalent of an abrupt climate change that would send us into Hot House conditions. However, only an additional 3.5 °C rise would be very bad news and probably result in some of the serious, possibly catastrophic, conditions discussed earlier. Under this scenario, sea level could certainly rise 1 meter or more in the next century with catastrophic effects.

Abrupt Climate Changes

The worst possible scenario would occur if we crossed a threshold that radically changes the world climate in a short period of time (a decade to a century). The chances of this happening are unknown, because we do not clearly understand the tripping mechanisms. However, if it did happen then civilization would be in jeopardy of collapse. An abrupt change could propel us into a Hot House by a possible catastrophic release of greenhouse gases. A catastrophic release of methane is not out of the question if the bottom waters in the northern oceans reach temperatures high enough to release the enormous quantities of methane locked up as methane hydrate in the bottom sediments (Leifer et al., 2006). This would lead to rapid melting, destabilization, and the possible collapse of large portions of the ice sheets raising the sea level by several meters.

The Worst Case

It is possible that global warming could bring about the end of civilization if we experience one of the more extreme rises in temperature. The survival of any civilization depends on crops, domestic animals, available water, natural resources, energy, a relatively stable society, and a viable economy. Crops provide grains, fruits, and vegetables to feed people and livestock. Domestic animals provide meat and dairy products supplemented by fishing to feed people. Civilization began because of agriculture and today civilization is completely dependent on it. How many of the 6.6 billion people could survive today on a hunter/gatherer existence? Water is also critical for irrigation and drinking, and natural resources for building, metal refining, industry, etc. Energy is required for heating, transportation, industry, etc. Also a fairly stable society and a healthy economy are required to keep things running smoothly. Climate change can disrupt all of these things either seriously or catastrophically. In fact, some ancient civilizations such as the Mayan and Anasazi are thought to have been destroyed by local climate change—in these cases, prolonged drought.

All of human evolution has taken place in an Ice House: alternating ice ages and interglacial periods. Our genetic code has evolved under Ice House conditions. Humans have never experienced Hot House conditions where there is no ice on Earth and torrid conditions occur everywhere on Earth except the extreme northern and southern latitudes. You might think that if humans can survive ice ages then surely they can survive global warming, but that is not necessarily the case. Although humans survived ice ages they were few in number and there was no civilization. The total human population during the ice ages was only a few million at most. As they were hunter–gathers, they did not require crops or domestic animals for their survival. When the ice sheets advanced they simply walked away from them. Civilization developed when the last ice age ended about 12,000 years ago, and the world enjoyed a more-or-less steady mild climate up to the present time. In fact, it is probably no coincidence that early *Homo sapiens* and modern *Homo sapiens* both developed during interglacial periods. The relatively equitable and stable climate allowed for agriculture and consequently the birth of civilization.

Today things are very different than they were just 100 years ago. There are about 6.6 billion people who are supported by the crops they grow, the domestic animals they use for food, and the technology that powers the global economy. This population is increasingly concentrated in urban centers. Furthermore, the environment is already stressed from overuse of land and resources and that is leading to the extinction of many species among other things. Global warming at its extremes could result in worldwide crop failures, mass starvation, political chaos, and economic collapse. The regions most susceptible to collapse are developing countries in Africa, South and Central America, and Asia that are already in precarious economic and political situations. Africa will be particularly vulnerable because 70% of the least-developed countries occur there. The people best suited to weather a collapse of civilization are those that do not depend upon it for their survival. They include primitive tribes in places like the Amazon jungle, the Arakan Yoma of Burma and India, and New Guinea, for example.

SUMMARY

No matter what we do, the global average temperature will probable climb about another 0.6-1 °C with serious consequences for us. If the CO_2 level reaches about 440 ppm then it will rise at least another 1.5 °C with severe consequences. At higher levels things get very serious and at the extremes of 4-6 °C they become catastrophic. We must start massive reductions of greenhouse gases now before it is too late. We must think ahead. The longer we delay, the greater the possibility of catastrophic consequences for our children and grandchildren.

13

THE ESCAPE HATCH

W E are as addicted to fossil fuels for our energy needs as the drug user is addicted to his favorite narcotic. We are now paying a heavy price for that dependence, and we must find a cure or we will end up in just as much trouble as an untreated drug addict.

Reducing greenhouse gas emissions is not as difficult as it may seem. However, it will take a major effort by the world community working in unison and in agreement to reduce emissions enough to stabilize the atmosphere. There are basically only two ways of reducing greenhouse gases: carbon sequestration and non-polluting energy sources. Each of these broad categories has numerous methods for either extracting emitted carbon and safely storing it, or using alternative energy sources. However, all of these methods have problems of one type or another. The best approach is to use as many methods as possible to lessen the burden on any one system. Only some of the possibilities are discussed below.

Carbon Sequestration

There are a number of ways to sequester CO_2, the most important greenhouse gas. Unfortunately, some of the methods of sequestering CO_2 can have deleterious effects on the environment. Probably the most commonly suggested method is to just bury it (Socolow, 2005).

CO_2 Burial

One method of burial is to capture the CO_2 and pump it to the bottom of the ocean where the very high pressures will keep it in liquid form and stuck on the sea floor for decades. It is estimated that the deep ocean can hold between 1,000 and 27,000 billion metric tons more than at present. The problem is that the oceans are already turning acidic from absorbing too much CO_2, and adding more carbon will mean that, some day, it will be so acidic that it may not support much marine life. Although this method has been tried under experimental conditions, there is no assurance that it would work in nature.

A better way is one that is already occurring in the North Sea where the Norwegian oil company Statoil is burying CO_2 in an aquifer below the seabed. The company is extracting CO_2 from natural gas wells and injecting it into an aquifer 900 meters below the surface. In this way the Sleipner West Gas field stores about 2,800 tons of CO_2 a day, which is about the same rate that CO_2 is produced in a 140-megawatt coal-fired power plant. The same method can be used in gas fields on continents. It could also be used to bury CO_2 emissions from power plants.

There are several different reservoirs that are suitable for storing CO_2 (Friedmann, 2003), the best of which are probably saline aquifers, which are porous and permeable strata that hold unusable brines. They are thought to be the largest reservoirs with an estimated potential of holding from 100 to 1,000 billion tons of CO_2. Before injection into the reservoirs the CO_2 must be converted to a high-density phase (supercritical) with the properties of both a liquid and a gas. The supercritical CO_2 displaces the pore fluids and is trapped below a seal called an aquiclude. On average, the depth must be below about 800 meters for long-term (>500 years) stability. One problem with this method is finding enough suitable reservoirs within a reasonable distance from the power sources. Transporting the huge amount of CO_2 required to significantly lower CO_2 emissions is an enormous problem.

Extraction from the Atmosphere

Another method of getting rid of CO_2 is by extracting it from the atmosphere and making it into carbonate rocks, as is done in nature. Geologists have been able to speed up a process that normally takes geologic time. They combined CO_2 with water to form carbonic acid that reacts with certain minerals, e.g. magnesium-rich serpentine, to create quartz and carbonates. They found that 80% of the magnesium silicate converted to carbonates in about 30 minutes.

A somewhat similar method of extracting CO_2 from the atmosphere has been proposed by scientists at the US Department of Energy's Los Alamos National Laboratory. In this method they propose to pass air over an extraction agent such as quicklime. The CO_2 in the air reacts with the quicklime to form calcium carbonate (limestone) that falls to the bottom of the extractor. According to the scientists, the limestone would be heated to form CO_2 and quicklime that is recycled back to the extractor. The CO_2 is then sequestered by burial or some other technique. To extract CO_2 emitted by humans requires an extractor equivalent in size to 1 m^2 for every person in the "industrialized countries," according to the scientists. However, the people in countries of the 10 largest emitters of CO_2 must be included to decrease human CO_2 emissions by 62%. Since there are a total of 3.2 billion people in these countries, the size of the extractor would have to cover an area of 3,200 km^2. However, there is no reason that thousands of smaller extractors could not be scattered around various countries of the world.

Another proposed way of extracting CO_2 directly from the atmosphere is a method called *ocean iron fertilization*. The idea is to add iron to the nutrient-rich Southern Ocean surface waters to increase the phytoplankton abundance and, therefore, increase the uptake of CO_2. Experiments to date show mixed results. It seems that you would have to fertilize something like 10^9 km^2 to absorb 30% of the carbon released by humans in one year. That is an area 10 times the area of the Southern Ocean. Furthermore, the phytoplankton abundance has already decreased an average of 8% due to warming of the oceans and will probably continue to decrease as the oceans warm further. The oceans have already absorbed 100 billion tons of human produced CO_2, and this has already resulted in changes of ocean chemistry, ecology, and climate. We probably should not be tampering any more with the oceans. Furthermore, recent studies have shown that iron fertilization is not capable of sequestering enough CO_2 to significantly lower the atmospheric abundance (Zeebe and Archer, 2005; Buesseler and Boyd, 2003). This method is just not feasible.

The ideal way to decrease the CO_2 content of the atmosphere is to extract it from the atmosphere. If we could decrease the concentration of CO_2 to the pre-industrial value then over time we could return to the temperatures of that time. If fact, we may be able to regulate the CO_2 content to some ideal value for humans and other species. This would certainly be difficult, but it could be the ideal way of solving the problem.

ALTERNATIVE ENERGY SOURCES

There are a number of non-polluting energy sources that can be used to generate electricity and fuel transportation. They include, but are not limited to, hydrogen fuel cells, wind power, nuclear power, hydroelectric power, solar cells, ocean currents, and geothermal power.

Hydrogen Fuel Cells

The transportation sector in industrialized nations is one of the major emitters of CO_2. In the United States the transportation sector is the second largest CO_2 emitter after power generation, and it is growing. Without addressing this sector it would be very difficult to significantly lower CO_2 emissions. Switching to hydrogen fuel cells for powering our cars and trucks would be a major way to reduce CO_2 emissions (Crabtree et al., 2004; Burns et al., 2002; Jacobson et al., 2005). In this technology the motor gets its power from a fuel cell unit. Electricity is produced when electrons are stripped from hydrogen atoms and travel through a membrane in the cell. The electric current runs an electric motor that turns the wheels. The hydrogen protons then combine with oxygen and electrons to form water. This type of fuel-cell car only emits water vapor and heat.

Hydrogen fuel cells have other major advantages over gasoline-fueled internal combustion engines. Today an internal combustion engine is only about 20 to 25% efficient in converting the energy content of gasoline into turning the wheels. That may improve to 30% in the future, but it probably will not go any higher. In contrast, the hydrogen fuel cell is up to twice as efficient at about 55%, so that it takes half the fuel energy to achieve the same driving power as an internal combustion engine. This efficiency will increase as membrane technology improves, allowing thinner membranes.

The growing world population and the growing affluence of developing nations means that personal transportation will grow substantially in the near future. Only 4% of the world's population had vehicles in 1960, but the percentage increased to 9% in 1980, and is 12% today (2006). By 2020 it will be 15% at the current growth rate. There are about 700 million vehicles in operation but that could climb to over 1 billion in the next 20 years. Today about 75% of all vehicles are found in the United States, Europe, and Japan, but in the future it is estimated that 60% of the increase in new car sales will be in China, Brazil, India, Korea, Russia, Mexico, Poland, and Thailand. In the United States alone about 65% of the imported oil is used to power the transportation system. If we are ever going to solve the CO_2 emission problem we will need to find some other form of energy to power our vehicles.

Producing hydrogen requires energy to reform hydrocarbons with catalysts, or electricity to split hydrogen from water. However, the high efficiency of fuel cells more than compensates for the energy required to produce the hydrogen. The ideal way would be to use renewable energy sources such as hydroelectric, solar, wind, or geothermal energy to produce hydrogen. The real problem is making hydrogen readily available to power fuel cell vehicles when that type of infrastructure is very difficult to build without the presence of a large number of fuel cell vehicles already on the road. Furthermore, hydrogen is extremely explosive and would be very difficult to manage in vehicles or transportation systems.

Wind Power

Wind turbines use rotational kinetic energy to drive electric power generators. The global kinetic energy of wind is estimated at about 11 quadrillion kilowatt-hours per year. Just 10% of this energy would far exceed the world's yearly demand of electricity. World wide there are about 25,000 wind turbines operating today, but they only produce about 0.1% of the world's electric power. Today wind turbines only convert about 40% of the wind's energy into useful energy. Furthermore, they depend on more-or-less constant winds to insure a reliable source of energy. It is unlikely that wind power will ever meet the power needs of the world, but it could be locally significant by providing whole communities with their power requirements if they are located at a proper site. Great Britain intends to significantly increase the generation of electricity by wind power by 2020.

Ocean Currents

One of the great undeveloped sources of power are ocean currents (Ehrenman, 2003), and Great Britain and Canada are developing ways of generating power from turbines using this method. These sea turbines work like wind turbines but derive their energy from coastal ocean currents. The great advantage of using water over wind is that seawater is 832 times denser than air, therefore, a 5-knot ocean current is the same as a wind velocity of 270 km/hr. Britain is testing a 300-kilowatt turbine that will be grouped in arrays producing 20 megawatts each. The turbine is connected to shore by a marine cable lying on the seabed. A Canadian study that proposed running 794 one-megawatt turbines over about 1,587 hectares would generate 1,390 billion watts per hour at a cost of 7 US cents per kilowatt-hour. The total cost of the project would be about $900 million.

Nuclear Power

Nuclear fission is a highly controversial power source because it produces highly toxic waste that must be disposed of in a safe place. Nuclear power plants can also have accidents that cause a great deal of harm, as the 1986 Chernobyl accident in Russia demonstrated. Opponents of nuclear power argue that it is costly, potentially dangerous, and vulnerable to terrorists attack; it is also difficult to dispose of the nuclear waste. Advocates say that nuclear power plants produce clean, cheap, and carbon-free electricity. Of course, if we could develop nuclear fusion power we would not have the toxic waste problem as that is the same way that energy is generated within the Sun and the end product is harmless helium.

Today nuclear energy produces only 6% of the world's commercial energy. In the United States it produces about 8% of the electricity, but in France it produces 80% of the electricity. A recent study of the future of nuclear power by the Massachusetts Institute of Technology (Ansolabe-here et al., 2003) asserts that nuclear power should be maintained as an energy option because it is a carbon-free power source. However, based on a survey of adults, the study observed that those who were very concerned about global warming are no more supportive of nuclear power than those who were not. Indications are that the European response would be about the same. The study recommended that the public be educated about the dangers of global warming and the need to use all available carbon-free power sources to curb greenhouse gas emissions. Currently Great Britain is re-evaluating its policy of phasing out nuclear-generating facilities by 2023. We need to think very carefully about the trade-offs of using nuclear energy against the possible consequences of global warming.

Hydroelectric Power

Worldwide hydroelectric power provides about 7% of the commercial energy, but only 3% in the United States. We might be able to build a few more dams, but it will certainly not be enough to meet all of our energy requirements. One encouraging development is the Three Gorges Dam in China that will provide electricity that is otherwise generated by coal-burning power plants.

Geothermal Power

Geothermal energy is derived from the Earth's internal heat. The temperature increases with increasing depth, and just a few kilometers deep the temperature can be over 250 °C. On average the temperature increases 1 °C every 36 meters in depth. In volcanic regions molten rock can be near the surface and provide easy access to tap its heat energy. The hot rocks heat water to produce steam that drives turbines producing electricity from electric generators. In the geothermal method, cold water is pumped down drill holes and heated to produce hot water and steam that emerges from another drill hole. The amount of energy generated depends on the temperature of the rocks and, therefore, the amount of water that is pumped down to the heat source. This is why volcanic areas are used to generate geothermal power. The steam must be purified before it is used to drive a turbine, otherwise the turbine blades will be ruined. Today geothermal energy is used to generate electricity in a number of volcanically active regions such as Iceland and New Zealand. In Iceland, geothermal heat is used to heat houses as well as to generate electricity.

The main advantages of geothermal energy are that it does not produce pollution or greenhouse gases, and no fuel is required. The main disadvantage is that there are not many places suitable for geothermal power stations. The hot rocks must be of a suitable type and at a relatively shallow depth. Also, hazardous gases may issue from the subsurface and be difficult to dispose of safely.

Solar Power

Various devices can change the Sun's energy into heat or electrical energy. Flat-plate collectors convert solar energy into heat energy to heat water and the air inside buildings, while solar cells convert solar energy into electric energy. Although solar power provides a clean unlimited source of energy, it is distributed over a wide area requiring vast arrays to capture and concentrate the energy. Also darkness and bad weather interrupt the supply of sunlight. Solar energy technology is developing very rapidly and it could prevent much emission of greenhouse gases if used by residences, commercial establishments, and other facilities.

GEO-ENGINEERING

There are other ways of cooling the Earth without reducing greenhouse gases. These are known as *geo-engineering* techniques. All of them reduce the amount of the Sun's radiation reaching the Earth's surface. One method proposes injecting sulfur into the stratosphere to deflect an amount of radiation sufficient to cool the Earth to some desired temperature. Another possibility is to artificially cool the Earth by putting trillions of small lenses into orbit at the inner Lagrange point to reflect some of the Sun's radiation back to space (Angel, 2006). Both of these geo-engineering methods have drawbacks: one produces too much acid rain, and the costs are enormous. Neither method addresses the fundamental cause of global warming—the emission of greenhouse gases. It is not clear why there is any advantage of using them instead of just reducing greenhouse gas emissions. In any case, they should only be used after thorough evaluation, and only as a last resort.

SUMMARY

We must use every means at our disposal to reduce greenhouse gas emissions. Time is running out. All of the methods of sequestering CO_2, and all of the alternative pollution-free energy sources should be used to reduce the atmospheric greenhouse gas content. Technology got us into this mess, and it can get us out. If enough resources are put into research on CO_2 sequestration and new energy sources, it is likely that we could avert a catastrophic warming, but we need to start implementing major emission reductions now.

14

POLITICS AND MONEY

S OLVING the problem of global warming requires a global response by all of the world's nations and particularly the largest emitters of greenhouse gases. The world's nations must agree to reduce their greenhouse gas emissions by about 60%. Otherwise we are heading for disaster.

THE KYOTO PROTOCOL

A first attempt at a major reduction in greenhouse gas emissions by the world community is the Kyoto Protocol. This agreement is only a first step to emissions reduction. It will not come close to solving the problem. It is, however, a major advance because the Protocol has the attention of the world. Unfortunately, the main emitters of greenhouse gases will do nothing, either because they did not ratify the Protocol or they are not required to reduce their emissions under the current rules. This is unacceptable.

The Kyoto Protocol is a pact that was agreed to by certain governments at a 1997 UN conference in Kyoto, Japan. The Protocol calls for the reduction of greenhouse gases emitted by developed countries. The average reduction is to be 5.2% below the 1990 level, beginning in the period 2008 to 2012. To date (2007), a total of 156 nations had ratified the pact. This is not the first international meeting on climate change. In 1992 governments agreed to address the problem of climate change at an "Earth Summit" in Rio de Janeiro. Kyoto was the follow-up meeting, and was the first legally binding global agreement to cut greenhouse gas emissions. The Kyoto Protocol went into legal force beginning February 16, 2005.

The penalty the Kyoto Protocol imposes on a country that misses its target is not very severe. For every ton of greenhouse gas it emits over its target from 2008 through 2012, the country must cut an additional 1.3 tons between 2013 and 2017. There is no stated means of enforcing the penalty.

Not all nations must cut their emissions by 5.2% below their 1990 level. Of the 156 nations that ratified the pact, only 39 relatively developed countries must meet the standard reductions because the more industrialized countries are the largest emitters. Under the Protocol rules four out of the five largest emitters do not have to cut their emissions (see Chapter 8): two (China and India) because they are considered developing nations; one (Russia) because its emissions are already 5.2% below their 1990 level; and one (United States) because it refuses to ratify the Protocol. President George W. Bush rejected the Kyoto Protocol in March 2001. He said it would harm the US economy by limiting the use of fossil fuels that were crucial to a healthy economy. Bush also opposed the protocol because it would not restrict emissions in developing countries, particularly those with very high emissions.

The current human-caused global emission rate of CO_2 is about 25 billion metric tons each year (31 if you include land use). If the Kyoto Protocol achieved a 5.2% reduction of global CO_2 emissions below the 1990 level, then the emission rate would be 21.42 billion metric tons per year (Figure 14.1). However, if it achieves a yearly reduction of 1 billion metric tons of CO_2 equivalent, with the non-participation of four of the largest emitters, it will be lucky. In fact, it appears that only Sweden and Great Britain will meet their 2012 reduction quotas. Even if the Kyoto Protocol goals were met it would still be woefully inadequate to do

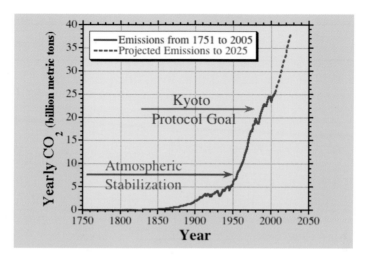

FIGURE 14.1 This graph shows the growth in the emission of CO_2 from 1750 to 2005, and its estimated growth from 2006 to 2025 (IEA World Energy Outlook, 2004). Also shown are the Kyoto Protocol goal of emission reduction and the reduction required to stabilize the atmospheric CO_2 content. It is obvious that the present Kyoto Protocol will not come close to solving the problem.

anything but barely slow down global warming; it would not come close to reducing emissions sufficiently to stabilize the atmosphere (Figure 14.1). The Kyoto Protocol does not take into account the emissions due to land use, decreasing carbon sinks, increased world population or increasing emissions of naturally occurring methane and CO_2 due to the warming process itself.

During the Siberian heat wave of 2003 wildfires burned forests with an area of about 22 million hectares—a region almost the size of the state of Oregon. The burning trees released about 917 million metric tons of CO_2 equivalent, which is roughly the same amount by which participating industrialized countries pledged to cut their emissions by 2012 under the Kyoto Protocol. Also, the human-caused release of the equivalent of 3 to 9.4 billion metric tons of CO_2 from the burning of peat and forests during the massive 1997 fires in Indonesia (Page et al., 2002) is considerably more than will be saved under the Kyoto Protocol. Nor does the Kyoto Protocol take into account the projected increase in the world's population by 1.85 billion during the next 25 years (see Chapter 9). Assuming that each new person only produces an extremely low 1 metric ton of CO_2 per year (the world average is 4 metric tons per year), there would be an additional increase of 1.85 billion metric tons per year by 2025. We have to reduce our emissions to levels much lower than those called for by the Kyoto Protocol. In order to stabilize the atmospheric greenhouse gas content we need about a 60% reduction of emissions.

Chapter 14

THE MONTREAL CLIMATE CONFERENCE

A meeting of the parties to the UN Framework Convention on Climate Change was held in Montreal, Canada, from November 28 to December 8, 2005. The purpose of the meeting was to negotiate an extension of the Kyoto Protocol beyond 2012. Unfortunately, the results of the meeting were disappointing. The United States refused to budge on its position of non-participation in the Kyoto protocol. If the United States refuses to accept binding agreements to reduce greenhouse gases, then why should China and India? The United States position is irresponsible, and it has completely isolated itself from the rest of the world. If the United States, China, Russia and India do not reduce their emissions, then the situation is hopeless and we had better start preparing for the consequences of global warming to come. Remember, the provisions of the Kyoto Protocol will not even come close to solving the problem.

ATTITUDES, DETERRENTS, AND REACTIONS

Religious, political, and moral beliefs apparently influence one's attitude toward global warming. Political conservatives and religious fundamentalists tend to be more skeptical about it than liberals or religious moderates. However, global warming does not care whether you are a right-wing conservative, a political moderate or a left-wing liberal. It does not care if you are a religious fundamentalist, a religious moderate, or an atheist. It also does not care if you are extremely rich or extremely poor. Global warming will eventually affect you no matter what your beliefs or financial situation. However, it will, in general, affect the poor sooner and more severely than the rich. But eventually it will adversely affect all of us.

An excellent book about the politics, reporting and a possible approach to implementing a solution to global warming is Ross Gelbspan's *Boiling Point* published by Basic Books (Gelbspan, 2004). It goes into much more detail on these aspects of the problem than are covered here, and I highly recommend it.

The media and most policy makers have no background in science and are often confused about how science works. Five years ago the media, most journalists, and most policy makers considered global warming a highly controversial subject. The fuel industry was treating global warming like the tobacco industry was treating the health hazards of smoking in the 1970s: by either denying it or ignoring it. That attitude is beginning to change in some circles, but certainly not all.

The Bush administration continues to resist any mandatory strategies to significantly reduce greenhouse gas emissions. It has tampered with government reports to change the perceived significance of global warming. As a scientist, I find this absolutely unacceptable; it is fraud at the most basic and crass level. A person who was chief of staff of the Bush White House Council on Environmental Quality edited out the well-proven conclusions on global warming agreed to by the majority of scientists in the draft reports of the US Climate Change Science Program (Figure 14.2). This is the agency responsible for coordinating research on global warming from 13 federal agencies. The White House council was a former lobbyist and "climate team leader" for the American Petroleum Institute. He is a lawyer with absolutely no training in science. When he was exposed by *The New York Times* he resigned his White House post to work for the ExxonMobil Oil Company. This type of behavior by any administration is completely unacceptable.

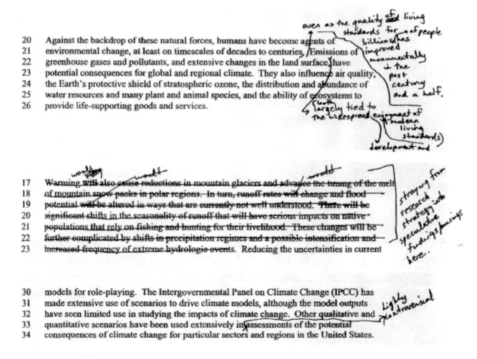

FIGURE 14.2 Excepts from the US Climate Science Program that have been edited by President Bush's former White House Council on Environmental Quality in 2005. The editor had no science background and now works for ExxonMobil Oil Company.

In June of 2005 Texas Republican Representative Joe Barton sent a letter to three scientists whose findings show that global temperatures have increased dramatically. The letter requested all the raw data that contributed to their research, all of their sources of research funding, all the papers they have published, and much more. Barton has also requested that the National Science Foundation provide a list of "all grants and other funding" given for climate research. This is clearly harassment of scientists whose research has been peer reviewed. Who is this Representative? Barton, who will chair the House-Senate conference on the energy bill, has close ties to the energy industry. Since 1987, he has received $1.84 million from the oil, gas, coal, nuclear, electricity and chemical industries—more than any other member of the House. The request by Barton was strongly criticized by the president of the University of Arizona where one of the scientists is employed, and by Senator John McCain, the senator from Arizona. It was also condemned by numerous scientists and their science societies.

Unfortunately, this type of interference continues. In early 2006, James Hansen, the noted climatologist and director of NASA's Goddard Institute for Space Studies, was told by a Bush appointee to the NASA public relations department that he would have his lectures, papers, postings on the Goddard website, and requests for interviews reviewed in order to stop him from speaking out about global warming. In video and telephone conferences two Bush political appointees that head the public affairs division said that more control of scientists' access to the media was needed, not less. NASA had better make some drastic changes in how they run the public affairs division. A good start would be for Congress to forbid the political appointment of anybody to governmental agencies that deal with science.

Similar types of suppression and harassment are occurring at the US National Oceanic and Atmospheric Administration, where administration officials have chastised scientists for speaking on policy questions; removed references to global warming from their reports, news releases and conference websites; investigated news leaks; and sometimes urged them to stop speaking to the media altogether. There is a real danger that Americans are not getting the full story on global warming, and that is reflected in recent polls discussed in the next chapter. The only way they could get a comprehensive picture of what is really happening to the climate is for them to read the science literature. Unfortunately, that is beyond their expertise, which is the primary reason I wrote this book.

Suppression of climate information is not confined to the United States. In Australia, the other country that did not ratify the Kyoto Protocol, three scientists studying climate change for the Commonwealth Scientific and Industrial Research Organization said that the government intimidated them over their views calling for the reduction of greenhouse gases. Two of the scientists, according to a Sydney newspaper, said they had been barred from disseminating scientific information on global warming.

This type of harassment and interference must stop. It is unethical. It violates free speech. It is disgusting, and it is dangerous.

Until recently the Bush administration refused to acknowledge the seriousness of the problem. When the IPCC 4th Assessment Report was released on February 2, 2007, President Bush, after six years of denial, finally admitted that global warming was a serious problem caused primarily by human burning of fossil fuels. However, unlike the European Union, he has still (April 2007) not proposed any laws to curb the emission of greenhouse gases.

IS THE TIDE TURNING?

Fortunately, there has been a recent shift in the debate over global warming outside the Bush administration. The media are now beginning to cover global warming more than in the past, but it is still a pathetically small coverage, particularly in the United States. The foreign media, such as the BBC, NHK (Japan) and many other major broadcasters, cover various aspects of global warming much more frequently than media in the United States. That is why the people of those nations have a much better understanding of the problem than Americans. The media in the United States are not doing their job. They are missing the greatest story since civilization began: a potentially catastrophic man–made climate shift. What more do they want, or are they influenced by their owners?

States and cities are beginning to take the initiative in reducing their own greenhouse gas emissions. In June 2005, the mayor of Seattle, Washington (Greg Nickels) began a campaign to lower greenhouse gas emissions on a city-by-city basis. Cities that agree to the pact will reduce their greenhouse gas emissions several percent below the Kyoto Protocol. In 2005, 174 cities across the nation had agreed to participate.

At the state level, 28 have agreed to develop strategies to reduce greenhouse gas emissions. California has passed legislation curbing greenhouse gas emissions much stronger than any other state. The legislation calls for a 25% reduction in CO_2 emissions by 2020. California has already (2004) adopted regulations to limit greenhouse gas emissions from cars. This goes into effect on the 2009 models. Since this requires automakers to re-engineer their cars and trucks, it should have benefits throughout the U.S. (It is unlikely they would only re-engineer their cars for the California market.) This can only be viewed as a start to a worldwide solution, but it is an important step because California emits more greenhouse gases than France.

The November 7, 2006 mid-term election in the United States put the Democrats in control of both the House and Senate. This makes it more likely—but very far from certain—that federal legislation could be passed curbing greenhouse gas emissions. Democratic Senator Barbara Boxer will now head the U.S. Senate's environmental committee and she has promised to make major shifts in U.S. global warming policy.

Speaking at a panel on January 18, 2006, commemorating the 35th anniversary of the US Environmental Protection Agency, six former heads of that agency, including five who served Republican presidents, said that the Bush administration needed to act more aggressively to curb the emission of greenhouse gases. All agreed that global warming was occurring and that humans were the primary cause. One former agency head under Richard Nixon, Russel E. Train, said "To sit back and push this away and deal with it sometime down the road is dishonest and self-destructive." At least somebody appears to be listening to the scientists.

The British Royal Society, the world's oldest scientific society, and two U.S. Senators, one Republican and the other a Democrat, urged ExxonMobil oil company to stop funding groups that spread the word that global warming is a myth or not primarily caused by burning fossil fuels. According to the Union of Concerned Scientists, ExxonMobile has funded 29 climate change denial groups in 2004 alone. The company has spent more than $19 million funding groups like the Competitive Enterprise Institute and the Tech Central Station Web site that promote their denial views through publications and Web sites that are not peer reviewed by the science community.

One of the most encouraging developments is that several US States— California, New York and eight others—are suing (April 2006) the Environmental Protection Agency for refusing to regulate greenhouse gas emissions from power plants, the biggest emitters of these gases. The suit tries to force the agency to start regulating emissions of greenhouse gases from power plants. The States' argument is that the agency has not lived up to the requirements of the Clean Air Act to ensure public safety. This is an extremely important litigation because 135 new coal-powered energy-generating plants are due to be constructed in the future. On April 2, 2007, the U.S. Supreme Court found that the EPA had the authority to regulate CO_2 as a pollutant and ordered the EPA to consider limiting the global warming pollution produced by motor vehicles.

The European Union is light years ahead of the United States and other major greenhouse gas emitters. In March 2007 they passed rules that over the next 13 years would cut greenhouse gas emissions at least 20% below the 1990 levels, produce 20% of their power through renewable energy, and see that one-tenth of all cars and trucks in the 27 EU nations should run on biofuels from plants. Britain went beyond the European Union rules by setting a target of 26% to 32% reduction during the same time-frame. It also proposed binding laws that would cut greenhouse gas emissions 60% by 2050.

A new movie and book called *An Inconvenient Truth* may have an impact on people's thinking about global warming. It was released nationally on June 16, 2006. It quite accurately and dramatically shows how global warming could affect the world and that it is real and caused by human activity. There have also been several books on global warming that could have a similar effect, hopefully including this one. If these books and movies can persuade people that we have a serious problem, then they will influence policy makers to take action to mitigate global warming.

CAN WE STOP GLOBAL WARMING?

The answer to that question is "not completely." There is going to be some further warming no matter what we do. But we can keep it from getting out-of-hand if we take immediate action. Remember, we have to prevent the atmospheric content of CO_2 from reaching about 440 ppm, at which point the global average temperature will inevitably climb to about 1.5 °C above today's value with probably serious consequences. We have less than about 30 years to accomplish this at today's rate of CO_2 increase. Some scientists think we only have about 10–15 years. Even if we do limit the CO_2 content to less than 440 ppm, glaciers will continue to melt and sea level will continue to rise. Therefore, we need to start drastically reducing emissions now if we are to protect our children's future. We cannot rely on voluntary emission reductions, nor can we rely on only a limited number of nations participating in emission reductions.

We have to get tough now. Global warming is a world problem requiring the action of all nations. They need to cut greenhouse gas emissions by 60% and they need to start now. Emission reduction should not be restricted to industrial nations. It should be mandatory for all nations, but proportional to the amount of greenhouse gases the nation emits. That burden will fall mostly on the top 10 emitters that account for 70% of the world's greenhouse gas emissions. The United States would be required to cut the most, followed by China, Russia, Japan, India, Germany, Canada, Britain, South Korea, and Italy. This does not mean that no other country will participate. It just means that countries with the least emissions will have the smallest reductions. How nations accomplish the reductions is completely up to them. Furthermore, there must be penalties for those nations that do not meet their required cuts otherwise there is no incentive to join in the reductions. Those penalties should be more severe for nations that refuse to take part in the plan. The penalties could take the form of trade sanctions of various forms or even prosecution by the World Court. Those countries that import fossil fuels could have those imports reduced, among other things, while those that export fossil fuels could be subject to more severe trade sanctions, such as discontinuing trade in some products important to their economy. On the other hand, countries that develop alternative energy devices (fuel cells, for example), sequestering techniques, devices for extracting CO_2 from the atmosphere or some other emission reduction device, and use and distribute these devices, should be acknowledged with perhaps financial rewards of some kind or favored trade status. This could provide incentives to develop new and innovative energy-saving or greenhouse gas reduction techniques that could generate large economic returns by developing whole new industries. This may sound like a very harsh method for reducing greenhouse gases, but it is about the only way to accomplish the required reductions in the shortest possible time. The alternative is continued global warming and a catastrophic future for our children and grandchildren. The United Nations need to set up some kind of organization to begin establishing the emission reductions for various nations, the penalties for non-compliance, and the incentives and rewards for innovations. The new organization could take the form of drastic changes in the Kyoto Protocol rules.

SUMMARY

The Kyoto Protocol, as presently structured, will not lower greenhouse gas emissions enough to stabilize the atmospheric CO_2 concentration, particularly because four out of the five largest emitters are not participating. United States government-appointed bureaucrats have either suppressed or altered global warming reports or research from scientists, which has led to confusion among the press and the public. However, the media and some politicians are beginning to realize that we have a serious problem, and some states are taking steps to curb greenhouse gas emissions independent of the federal government. We still have a very long way to go before the world governments, particularly the United States, understand the magnitude of the global warming problem and the urgency of taking strong steps to curb greenhouse gas emissions.

15

IT'S NOW OR NEVER

BY now you should clearly understand that we have an extremely serious problem that requires immediate action by the global community. It is unprecedented since civilization began, and can have catastrophic consequences for us and other species if we do not act soon. We are like children playing with matches, and we have already started a fire. Will we put it out, or will we just stand there and watch the house burn down?

There was a television commercial run by the American Dairy Association in the 1960s to combat the falling sales of butter because of the rising sale of non-dairy margarine. It opened with a man spreading margarine on a slice of bread. Next appears Mother Nature in a long flowing gown, a tiara, and a magic wand. She says, "You mustn't fool Mother Nature." She then zaps the man with her wand and turns him into a frog. Well, we have been fooling Mother Nature for 200 years and now she is getting angry. She is beginning to show her anger by adversely affecting the environment and if we continue to "fool" Mother Nature she will surely "zap" us with extremely unpleasant events. Getting rid of her irritant is one way Mother Nature solves her problems.

Do not worry about the Earth. It has suffered far greater traumas than climate changes and survived, but with major and lasting changes. Earth will continue to survive at least another 6 billion years before the expanding Sun engulfs it. The main worry is the future of humans, as we may be the cause of our own demise. As Walt Kelly's cartoon character *Pogo* said, "We have met the enemy, and he is us."

Keep in mind that the current consequences of global warming discussed in previous chapters are the result of a global average temperature increase of only $0.5\,^\circ C$ above the 1951–1980 average, and these consequences are beginning to accelerate. Think about what is in store for us when the average global temperature is $1\,^\circ C$ higher than today. That is already in the pipeline, and there is nothing we can do to prevent it. We can only plan strategies for dealing with the expected consequences, and reduce our greenhouse gas emissions by about 60% as soon as possible to ensure that we don't experience even higher temperatures. There is also the danger of eventually triggering an abrupt climate change that would accelerate global warming to a catastrophic level in a short period of time. If that were to happen we would not stand a chance. Even if that possibility had only a 1% chance of occurring, the consequences are so dire that it would be insane not to act. Clearly we cannot afford to delay taking action by waiting for additional research to more clearly define what awaits us. The time for action is now.

In June 2005 there was an ABC News poll of Americans concerning global warming. In that poll 59% were convinced that global warming was under way, but only 33% thought it would affect their lives. Only 38% favored immediate government action to deal with the problem. A majority (67%) thought that global warming posed no serious threat in their own lifetime, but 79% thought it posed a "serious threat to future generations." A more recent Gallup poll in April 2006 gave similar results and found that Americans were more worried about water pollution, air pollution (not greenhouse gases), and toxic waste than they were about global warming.

What are these polls telling us about American attitudes concerning global warming? First, the majority believes that global warming is happening, but it is only a small majority. Apparently 41% think global warming is not happening at all, although 99.9% of the climate scientists, including all major science academies, are convinced that global warming is real.

The fact that only a third of the people thought global warming would affect their lives suggests that the other two-thirds think it is so far in the future that it won't affect them. During the summer of 2003 when the European heat wave killed over 30,000 people, I gave a lecture in England on global warming. I said that if you were over 70 it was unlikely that you would be greatly affected by global warming. To my chagrin when I returned home, I learned that over 30,000 Europeans died from the heat wave and that 80% were over 75 years old. Needless to say, I have never repeated that statement. In fact the elderly are the most vulnerable to global warming, because they are least able to cope with the problems it causes them. Certainly the children alive today will be greatly affected by global warming and even those in their middle age will probably see significant and disastrous affects before they die. In fact, it could be the cause of their death. Future generations may look back at us with complete disgust because we knew what was happening but did nothing to prevent it.

The fact that a large majority (79%) thought global warming posed a "serious threat to future generations" is puzzling, particularly since over 40% did not think it was even happening. Either they think that "future generations" means generations in the distant future that are not yet born, or they do not care about the future of their children or grandchildren. Even if they think "future generations" means in the distant future, they apparently do not care what happens to their descendants. This is truly appalling. The most discouraging finding is that the majority do not think global warming needs our immediate attention.

What this poll indicates to me is that the majority of Americans do not understand the problem or the time scale for warming. They are mostly uninformed or misinformed about global warming, and have been misled by the government and the media. Fortunately, not all the general public are so ill informed.

I had a telephone call from a woman who was very concerned about global warming and wanted to know what she could do to help to solve the problem. I told her several things that would lower her emissions of greenhouse gases, and advised her to contact her congressmen, urging them to act now to curb greenhouse gas emissions. I asked her age and she said she was 83 years old. I told her that she did not have to worry about global warming because she would probably not live long enough to be affected by the problem. She said, " Oh, I'm not worried about myself. I'm worried about my grandchildren." I might have known that would be the response from part of the "Greatest Generation."[1] It is a shame that

---

[1] The "Greatest Generation" is the one that survived the Great Depression in the 1930s, and fought and won the Second World War.

most American journalists, the media, and particularly policy makers do not have the same attitude as one of the generation that saved them from a horribly repressive future. Are we now being led by the "Worst Generation"?

The public ignorance of global warming has been fostered by the Bush administration that refuses to do anything concrete about the problem. Washington, DC, apparently still does not understand the threat posed by global warming. Other countries are at least partly aware of the seriousness of the problem and have urged immediate action to curb greenhouse gases. A notable example is the European Union that is instituting strong measures to curb greenhouse gas emissions. It would seem that America's policy makers do not trust the world science community or even their own scientists. That is probably because most policy makers are not trained in science, or they do not care about the future world condition.

Today the world is pouring vast amounts of money into stopping global terrorism, but to date we have done very little to curb global warming or to prepare for its consequences. Since 2001, terrorist attacks in places like New York, Madrid, Istanbul, Indonesia, and London have killed several thousand people. However, in Europe during two weeks in the summer of 2003, 30,000 people were killed by a heat wave greater than any recorded in at least the past 500 years. In other words, 30,000 people were killed by one of the consequences of global warming: the increase in the frequency and intensity of extreme weather events. You can imagine the world reaction if 30,000 innocent Europeans were blown to pieces by terrorists in two weeks. Most people just do not understand the seriousness of the problem—including the government of the largest emitter of greenhouse gases. If something is not done now to address the climate problem, then, in the not too distant future, terrorism will seem like a tea party compared to the loss of human life from global warming. Even the CIA warned that global warming will stimulate an increase in conflicts over decreasing water and food supplies. It proposed seven responses to this threat, but not a single one involved curbing greenhouse gas emissions to mitigate global warming.

It may take a major catastrophe that is obviously the result of global warming before people will demand that we do something to curb it. Unfortunately, by the time something like that happens it will probably be impossible to stop. All we can do at that point is try to adapt to the changed climatic conditions, which may not be possible for much of the world.

There are several aspects of the current climate change that are deeply disturbing: (1) the abundance of greenhouse gases and their rates of increase are greater than at any time in at least the past 650,000 years; (2) the rate of CO_2 growth is rapidly increasing; (3) the current rate of temperature rise is greater than at any time during the past 10,000 years; and (4) some CO_2 sinks are decreasing and important sinks are now starting to become greenhouse gas sources. As we enter new territory in which we have no experience, predicting the future consequences of current global warming is a very daunting task. However, there is no question that the consequences will be very serious and possibly catastrophic. Today we face the greatest threat to humanity since civilization began, and for the reasons given above there is great urgency to immediately begin vigorously to reduce greenhouse gases. We will have to work very hard to prevent the global temperature from reaching critical levels. To date we have been as effective at curbing global warming as the old League of Nations was at preventing the Second World War.

In Chapter 1 the probability of solving the problem was given a 0–10% chance. You may think that it is a very pessimistic outlook or an alarmist view, but it is realistic at the present time, based on the world's current effort to reduce CO_2 emissions. There are two sets of curves that will indicate progress in solving the global warming problem. One is the global human–caused CO_2 emissions graph shown in Figure 7.8, and the other is the Keeling curves in Figures 7.5 and 7.6, showing the rising atmospheric CO_2 content and the rate of CO_2 increase. The first sign of progress is when the human emissions curve begins a steady decrease independent of the world economy (economic recessions result in decreases of human emissions). At that point the probability of a solution is about 50%. However, by far the most important curve is the Keeling curve that records the increase in atmospheric CO_2. When the rate of increase of CO_2 emissions (Figure 7.6) slows from its current ~ 2 ppm for a significant period of time (maybe about five years) then real progress has been made, and the probability of eventually stabilizing the atmospheric CO_2 content might rise to about 70–90%. Of course, by that time there should be a significant decrease in human greenhouse gas emissions. When the atmospheric CO_2 content (Figure 7.5) flattens out, we have then had a significant effect on the climate. Of course, this could take over a century due to the inertia of the climate system discussed earlier. On the other hand, if the rate of increase rapidly escalates for more than about five years independent of El Niño events, then we have probably crossed a threshold where CO_2 emitted from natural sources is out of control and/ or CO_2 sinks have significantly decreased. At this point the probability of a solution is zero. Even if we are successful in stabilizing the atmospheric CO_2 content, we still must adapt to the consequences of a warmer planet.

One way of looking at our problem is by considering a fast-moving car that has a lot of momentum, like the climate system. When you take your foot off the gas pedal the car keeps moving until it eventually comes to a stop. Similarly, when greenhouse gas emissions are halted the climate continues to warm until it reaches some equilibrium value.

Imagine a family driving down a highway at 70 miles per hour in a car without brakes (the current climate system). In the back seat are children too small to see out the front window (our grandchildren). The accelerator in this car operates as usual but the power source is greenhouse gases instead of gasoline. The soft-spoken wife (science community) of the driver (world governments) warns him that, in the distance, there is a concrete barrier across the highway and he should take his foot off the gas (reduce greenhouse gas emission by 60%) and coast to a stop before it is too late. He has forgotten his glasses and does not see the barrier. Also, there is a bee in the car that is "terrorizing" him, distracting him from his driving, and he continues to drive forward at increasing speed.

This is the situation we are in today. The fate of the family is in the hands of the driver. It all depends on how he responds to his wife's warning. If he would have listened to his wife sooner by taking his foot off the accelerator earlier he would have coasted to a stop before hitting the barrier. That is no longer possible. He has waited too long. If he takes his foot off the gas pedal now (60% reduction in greenhouse gas emissions) the car's momentum will still be enough to send him into the barrier, but at a speed low enough to hopefully cause only minor damage. If he keeps his foot on the pedal with the same pressure (no increase or decrease in the amount of greenhouse gas emission) he would not accelerate. However, when he finally takes his foot off the gas pedal, the car will be so close to the barrier that he will hit it at high speed with disastrous results. Unfortunately, our driver is pushing on the accelerator (releasing an increasing abundance of greenhouse gases), going faster and faster, and will hit the barrier sooner and at higher speed. When he gets close enough to see the barrier (becomes aware of the seriousness of the problem) it will be too late. At that point, he will be going so fast and be so near the barrier that even if he took his foot off the gas pedal he would still crash into the barrier at a very high speed with catastrophic results for his family. This is the reason we must act now to reduce greenhouse gas emissions. The longer we wait, the more difficult it becomes to prevent a calamity.

Global warming is real, it is primarily caused by the emission of greenhouse gases by human activity, and it will have serious—possibly catastrophic—consequences if we do nothing to mitigate the problem. The time to act is now. Let's take our foot off the accelerator before it's too late.

APPENDIX A

MEASURING GEOLOGIC TIME

Geologic time is simply the age of the Earth and, therefore, the age of the Solar System. One of the great triumphs of modern science is the ability to determine the age of rocks. This has enabled us to establish a chronology of Earth's history, determine the ages of extraterrestrial objects such as asteroids, and, indeed, calculate the age of the Solar System.

The technique for measuring the age of a rock is called *absolute* or *radiometric age dating*. Certain isotopes are radioactive and undergo spontaneous changes that affect the number of protons and neutrons in their unstable nucleus. The unstable *parent isotope* is changed into a stable *daughter isotope*. Over time the unstable parent isotope decreases in abundance and the stable daughter isotope increases in corresponding numbers. The rate at which a radioactive isotope decays is a unique property of each isotope and is called the *decay constant*. The time taken to reduce the number of parent isotopes by one half is called the *half-life* of the isotope. Since the decay rate is known, one simply measures the number of parent and daughter isotopes and calculates the age.

Radiometric age dating involves determining the time since a radioactive isotope becomes fixed in a system such as crystals in rocks. This occurs when rocks solidify from a molten condition. In the liquid or gaseous state, the parent or daughter atom can escape from the system. As a rock solidifies from a melt and begins to crystallize, some minerals attract a certain parent element but not the daughter product. As soon as these crystals solidify from a melt, the radiometric clock starts ticking. Therefore, the radiometric age of a rock is the age that it solidified from a melt.

Only igneous rocks that solidified from molten material can be used to determine absolute ages. Sedimentary rocks are composed of the eroded remains of igneous rocks and other material so they cannot be age-dated with any accuracy. Such determinations would give the solidification age of the crystals that make up the sediment, not the time the sediment was deposited. However, the sedimentary record is interspersed with lavas and other types of igneous rocks that allow us to infer the absolute ages of sediments.

Isotopes used in age dating must have a long half-life in order to determine the ages of very old rocks. Conversely, very young material must use isotopes with short half-lives to get reliable results. For young material, not enough time has elapsed for a significant accumulation of daughter isotopes from parent isotopes with long half-lives. For example, rubidium 87 decays to strontium 87 with a half-life of 49 billion years. These and other long half-life isotopes are used to measure very old igneous rocks. Carbon 14 is used for measuring young ages of archeological importance, such as ancient fires, and other organic remains that contain carbon 14. Neutrons produced in the upper atmosphere by cosmic radiation react with ^{14}N to form the radiometric isotope ^{14}C. Carbon 14 is concentrated in organic matter and decays back to ^{14}N with a half-life of 5,730 years.

The *minimum* age of the Solar System is the oldest solidification age measured. The oldest Earth material found to date (zircons) is 4.4 billion years. The oldest known lunar rock is 4.2 billion years, and the oldest known Martian rock (a meteorite) is also 4.2 billion years. Most meteorites are pieces of asteroids that are thought to have formed at the very beginning of Solar System formation. The oldest meteorites from asteroids cluster around 4.5 billion years. Complex measurements of various stable and unstable isotopes (^{204}Pb, ^{86}Sr, etc.) allow estimates of the time when the initial isotopic mixture of a rock's parent material was established. In other words, this is an estimate of the date of formation of original planetary material. Using this method, terrestrial, lunar, and meteorite samples all converge on an age of 4.54 ± 0.04 billion years. This is the age of the Solar System.

APPENDIX B

CLIMATE PROXIES

How do we know what the climate conditions were in the geological past? There were no meteorological stations, no thermometers, no barometers or any other weather instruments. Reliable worldwide weather record keeping only began in 1880. Today modern science has allowed us to reconstruct the climate conditions during much of Earth history by measuring and studying certain characteristics of ancient and recent sediments, ice cores, tree rings, corals, and other features. These characteristics are called *climate proxies*. Obviously the recent past (the last few million years) provides the best information, because there is much more data for them than for older sediments. There are still many ancient deposits to be examined and much more to be learned about past climates.

The climate proxies we have studied so far have yielded important information on past climates and their causes (*paleoclimatic and paleoceanographic data*). Most proxy measurements are only available from limited locations, and that makes it difficult to draw conclusions about global climate. Measurements of the same age from widely separated locations—for example, Greenland ice sheet, mid-latitude mountain glaciers and the Antarctic ice sheet—provide the most reliable information on past global climates. Also, ancient ocean sediments provide sufficient data to decipher climate and ocean conditions approaching a global scale. Determining past climate conditions depends on the use of a variety of proxies rather than just one. A few of the most common proxies are discussed below.

Glacial Sediments

Glaciers and ice sheets deposit distinct types of sediment that sporadically occur throughout the geologic record. These deposits consist of *tillites* and *stratified glacial drift*. *Till* is deposited as glacial ice melts and drops its load of rock fragments. Stratified drift is deposited by glacial meltwater, and is sorted according to the size and weight of the fragments. Striated boulders are often found in the tillite deposits where they have been scoured by abrasion at the base of the glacier. They are called *glacial erratics*. These sedimentary deposits have distinct glacial characteristics that distinguish them from other origins.

Other glacial landforms such as *moraines, outwash plains, eskers* and *kettles* are found in more recent glacial deposits. Moraines are layers or ridges of till, while outwash plains are sediments deposited at the end of a glacier as it melts. Eskers are ridges formed by stream sediments deposited within tunnels in glaciers, and kettles are large depressions caused by the melting of embedded masses of ice detached from the main glacier. Extensive glacial deposits have been found in various ancient sediments such as in the Permian, 250 million years ago, and near the end of the Precambrian, 600 million years ago. These deposits indicate ice ages and climatic changes millions of years in the past.

Stable Isotope Ratios

There are a number of stable isotopes, each having different atomic weights. They are used to estimate past climates, and ocean and air temperatures. Most chemical elements have several different stable isotopes. For example, most oxygen atoms have 8 protons and 8 neutrons for a total atomic weight of 16 (^{16}O), but a few have 10 neutrons and 8 protons for an atomic weight of 18 (^{18}O). Some even have 9 neutrons (^{17}O).

Any isotope of a particular element can form a compound with other elements. Water (H_2O), for example, can be made with ^{16}O, ^{17}O, or ^{18}O, and it can also be made with regular hydrogen (^1H), deuterium (^2H), or with radioactive tritium (^3H). It is differences in the ratio of ^{18}O and ^{16}O that permit the determination of temperature and climate. This difference in ratios is called *isotopic fractionation*—the sorting of isotopes by weight in different compounds. A variety of stable isotopes are used to determine other characteristics of the climate.

The abundance of ^{18}O increases in the ocean when ice sheets form on the continents. The atmosphere, together with rain and snow, is enriched in ^{16}O relative to the ocean. When the ice forms it captures ^{16}O preventing it from reaching the ocean in rivers. This makes the ocean even more abundant in ^{18}O than it already was. By measuring the ratio of ^{16}O to ^{18}O in an ancient shells or other objects we can learn how much ice was present when the shell formed. Furthermore, the extent of oxygen isotope fractionation is dependent on temperature: the warmer the temperature the less fractionation. For example, shells formed in cold water will have more ^{18}O than those formed in warm water. This means that past temperatures can be determined by measuring stable isotopic concentrations in a variety of materials including ice, calcium carbonates (limestones and shells), sediments, seawater, trees, bones, and many other materials. Other stable isotopes, such as those of carbon and hydrogen, are also used to determine various aspects of past climates and atmospheric gas abundances.

Ice Cores

The climate record for the past 650,000 years has been obtained from ice cores from the Greenland and Antarctic ice sheets, and from mountain glaciers at various latitudes. The isotopic composition of ^{16}O and ^{18}O in water molecules is used to determine past temperatures and snow accumulation. Small air bubbles trapped in the ice provide a direct sample of the amount of each gas present in the atmosphere when the snow fell. This gives a history of the fluctuations of greenhouse gases such as carbon dioxide (CO_2) and methane (CH_4), and how they may have contributed to past climate changes. The dust content of the ice provides a measure of atmospheric circulation. Changes in the amount of dust correlate with shifts in the temperature, indicating changes in atmospheric circulation patterns. The acidity of the ice layers provides a clear indication of major volcanic eruptions. Large volcanic eruptions inject large quantities of sulfur compounds in the stratosphere to produce sulfuric acid aerosols that are precipitated in the snow. This is a good indication of the climate impact of past eruptions.

Ocean Sediments

Deep-sea sediments provide some of the best evidence for past global climate conditions. A substantial ocean drilling program over the past 30 years has provided numerous sediment cores from a variety of locations. Isotope ratios from marine shells that occur near the ocean surface and near its bottom have provided the ocean temperature variations with depth and over time. They also indicate the size of ice sheets at high latitudes that indirectly indicates fluctuations in global temperatures. Lake sediments can also be used for similar purposes.

Other Proxies

There are many other proxies that indicate climate conditions in the past. Tree rings, pollen, coral growth bands, bore hole temperatures, warm and cold water organisms, and stalactites provide climate information of one type or another, including wet and dry periods, sea level changes, and water temperatures. It is the combined measurements of these climate proxies that give us a picture of past climate changes. As more measurements are made the accuracy and extent of past climate changes will be much better understood.

APPENDIX C

STATEMENT ON GLOBAL WARMING

by the
**American Association
for the Advancement of Science**[1]
December 9, 2006

The scientific evidence is clear: global climate change caused by human activities is occurring now, and it is a growing threat to society. Accumulating data from across the globe reveal a wide array of effects: rapidly melting glaciers, destabilization of major ice sheets, increases in extreme weather, rising sea level, shifts in species ranges, and more. The pace of change and the evidence of harm have increased markedly over the last five years. The time to control greenhouse gas emissions is now.

The atmospheric concentration of carbon dioxide, a critical greenhouse gas, is higher than it has been for at least 650,000 years. The average temperature of the Earth is heading for levels not experienced for millions of years. Scientific predictions of the impacts of increasing atmospheric concentrations of greenhouse gases from fossil fuels and deforestation

---

[1] The **AAAS** is the world's largest science organization and publisher of the prestigious journal *Science*

match observed changes. As expected, intensification of droughts, heat waves, floods, wildfires, and severe storms is occurring, with a mounting toll on vulnerable ecosystems and societies. These events are early warning signs of even more devastating damage to come, some of which will be irreversible.

Delaying action to address climate change will increase the environmental and societal consequences as well as the costs. The longer we wait to tackle climate change, the harder and more expensive the task will be.

History provides many examples of society confronting grave threats by mobilizing knowledge and promoting innovation. We need an aggressive research, development and deployment effort to transform the existing and future energy systems of the world away from technologies that emit greenhouse gases. Developing clean energy technologies will provide economic opportunities and ensure future energy supplies.

In addition to rapidly reducing greenhouse gas emissions, it is essential that we develop strategies to adapt to ongoing changes and make communities more resilient to future changes.

The growing torrent of information presents a clear message: we are already experiencing global climate change. It is time to muster the political will for concerted action. Stronger leadership at all levels is needed. The time is now. We must rise to the challenge. We owe this to future generations.

APPENDIX D

CLIMATE-RELATED WEB SITES

http://www.realclimate.org
http://www.metoffice.gov.uk
http://www.ncdc.noaa.gov/oa/climate/globalwarming.html
http://www.ipcc.ch/
http://www.whrc.org/resources/online_publications/warming_earth/
    index.htm
http://data.giss.nasa.gov/gistemp/
http://www.climatehotmap.org/
http://maps.grida.no/
http://www.wmo.ch/index-en.html
http://illconsidered.blogspot.com/2006/03/guides-by-category.html
http://www.sierraclub.org/globalwarming/health/
http://www.globalwarmingart.com/wiki/Main_Page
http://earthobservatory.nasa.gov/Newsroom/

REFERENCES

Alley, R.T. Icing the North Atlantic, *Nature*, 392, 335–337, Mar. 1998.

Alley, R.B., J. Marotzke, W.D. Nordhaus, J.T. Overpeck, D.M. Peteet, R.A. Pielke Jr, R.T. Pierrehumbert, P.B. Rhines, T.F. Stocker, L.D. Talley and J.M. Wallace. Abrupt climate change, *Science*, 299, 2005–2010, Mar. 28, 2003.

Alley, R.B. Abrupt climate change, *Scientific American*, 62–69, Nov. 2004.

Alley, R.B., P.U. Clark, P. Huybrechts and I. Joughin. Ice-sheet and sea-level changes, *Science*, 310, Oct. 21, 2005.

Alverson, K., R. Bradley and T. Pederson. *Environmental Variability and Climate Change*, International Geosphere–Biosphere Program, 2001.

Andreae, M.O., C.D. Jones and P.M. Cox. Strong present-day aerosol cooling implies a hot future, *Nature*, 435, 1187–1190, 2005.

Andronova, N. and M. Schlesinger. Causes of global temperature changes during the 19th and 20th centuries, *Geophys. Res. Letters*, 27, Gl006109, 2000.

Andronova, N., M. Schlesinger and M.E. Mann. Are reconstructed pre-industrial temperatures consistent with instrumental hemisphere temperatures?, *Geophys. Res. Letters*, 31, L12202, 2004.

Angel, R. Feasibility of cooling the Earth with a cloud of small spacecraft near the inner Lagrange point (L1), *Proc. National Acad. Sci.*, 103, no. 46, 17184–17189, 14 Nov. 2006.

Ansolabehere, S., J. Deutch, M. Driscoll, P.E. Gray, J.P. Holden, P.L. Joskow, R.K. Lester, E.J. Moniz and N.E. Todreas. *The Future of Nuclear Power*, Mass. Instit. of Tech. Study, 2003.

Antonov, J.I., S. Levitus and T.P. Boyer. Thermosteric sea level rise, 1955–2003, *Geophys. Res. Letters*, 32, L12602, 2005.

Arctic Climate Impact Assessment. *Impact of a Warming Arctic: Arctic Climate Impact Assessment*, 144 pp., Cambridge Univ. Press, New York, 2004.

Arendt, A.A., K.A. Echelmeyer, W.D. Harrison, C.S. Lingle and V.B. Valentine. Rapid wastage of Alaska glaciers and their contribution to rising sea level, *Science*, 297, 382–385, Jul. 19, 2002.

Bard, E. Climate shock: Abrupt changes over millennial time scales, *Physics Today*, 55, 32–38, Dec. 2002.

Ballantyne, A.P., M. Lavine, T.J. Crowley, J. Liu and P.B. Baker. Meta-analysis of tropical surface temperatures during the Last Glacial Maximum, *Geophys. Res. Letters*, 32, L05712, 2005.

Barnett, T.P., D.W. Pierce and R. Schnur. Detection of anthropogenic climate change in the world's oceans, *Science*, 292, 270–274, Apr. 13, 2001.

Barnett, T.P., D.W. Pierce, K.A. AchutaRao, P.J. Gleckler, B.D. Santer, J.M. Gregory and W.M. Washington. Penetration of human-induced warming into the world's oceans, *Science*, 309, 284–287, Jul. 8, 2005.

Bauer, E., M. Claussen and V. Brovkin. Assessing climate forcing of the Earth system for the past millennium, *Geophys. Res. Letters*, 30, 9-1 to 9-4, 2003.

Beig, G and 18 co-authors. Review of mesospheric temperature trends, *Reviews of Geophys.* 41, 1-1 to 1-41, 2003.

Bellamy, P.H., P.J. Loveland, R.I. Bradley, R.M. Lark and G.J.D. Kirk. Carbon losses from all soils across England and Wales 1978–2003, *Nature*, 437, 245–248, 2005.

Bellouin, N., Boucher, O., Haywood, J. and Reddy, M., Global estimate of aerosol direct radiative forcing from satellite measurements, *Nature*, 438, 1138, Dec. 22/29, 2005.

Beltrami, H., J.E. Smerdon, H.J. Pollack and S. Huang. Continental heat gain in the global climate system, *Geophys. Res. Letters*, 29, 8-1 to 8-4, 2002.

Beltrami, H., G. Gosselin and J.C. Mareschal. Ground surface temperature in Canada: Spacial and temporal variability, *Geophys. Res. Letters*, 30, 6–1 to 6–4, 2003.

Beniston, M. The 2003 heat wave in Europe: A shape of things to come? An Analysis based on Swiss climatological data and model simulations, *Geophys. Res. Letters*, 31, L02202, 2004.

Bentley, C.R. If west sheet melts rapidly; what then?, *Geotimes*, 25, 20–21, 1980.

Berger, A., C. Tricot, H. Gallee and M.F. Loutre. Water vapour, CO_2 and insolation over the last glacial–interglacial cycles, *Philosophical Trans.: Biol. Sci.*, 341, 253–261, 1993.

Berthier, E., Y. Arnaud, D. Baratoux, C. Vincent and F. Remy. Recent rapid thinning of the "Mer de Glace" glacier derived from satellite optical images, *Geophys. Res. Letters*, 31, L17401, 2004.

Bice, K.L., D. Birgel, P.A. Meyers, K.A. Dahl, K-U, Hinrichs and R.D. Norris. A multiple proxy and model study of Cretaceous upper ocean temperatures and atmospheric CO_2 concentrations, *Paleoceanography*, 21, PA2002, 2006.

Bintanja, R., R.S.W. vande Wal and J. Oerlemans. Modelled atmospheric temperatures and global sea levels over the past million years, *Nature*, 437, 125–128, Sept. 1, 2005.

Bradley, R.S. and P.D. Jones (eds). *Climate Since AD 1500*, Routledge, London, UK, 1995.

Bradley, R.S., M. Vuille, H.E. Diaz and W. Vergara. Threats to water supplies in the tropical Andes, *Science*, 312, 1755–1756, Jun. 23, 2006.

Bradshaw, W.E. and C.M. Holzapfel. Evolutionary response to rapid climate change, *Science*, 312, 1477–1478, Jun. 9, 2006.

Brady, E.C., R.M. DeConto and S.L. Thompson. Deep water formation and poleward ocean heat transport in the warm climate extreme of the Cretaceous (80 Ma), *Geophys. Res. Letters*, 25, 4205–4208, Nov. 15, 1998.

Braganza, K., D.J. Karoly and J.M. Arblaster. Diurnal temperature range as an index of global climate change during the twentieth century, *Geophys. Res. Letters*, 31, L13217, 2004.

Brewer, P.G., C. Paull, E.T. Peltzer, W. Ussler, G. Rehder and G. Friederich. Measurements of the fate of gas hydrates during transit through the ocean water column, *Geophys. Res. Letters*, 29, 38-1 to 38-4, 2002.

Brown, G.C., C.J. Hawkesworth and R.C.L. Wilson (eds). *Understanding the Earth: A New Synthesis*, Cambridge University Press, Cambridge, UK, 1992.

Brown, L.R. *Plan B: Rescuing a Planet Under Stress and a Civilization in Trouble*, Earth Policy Institute, Washington, DC, 2003.

Brown, S., P. Stott and R. Clark, *Temperature Extremes, the Past and the Future*, Hadley Centre for Climate Prediction and Research, Exeter, UK, 2005.

Bryant, D., L. Burke, J. McManus and M. Spalding. *Reefs at Risk: A Map-Based Indicator of Threats to the World's Coral Reefs*, World Resources Institute, 56 pp., 1998.

Buesseler, K.O. and P.W. Boyd. Will ocean fertilization work?, Science, 300, 67–68, Apr. 4, 2003.

Burns, L.D., J.B. McCormick and C.E. Borroni-Bird. Hydrogen fuel-cell cars could be the catalyst for a cleaner tomorrow, *Sci. Amer.*, 65–73, Oct. 2002.

Burroughs, W.J. *Weather Cycles: Real or Imaginary?*, Cambridge University Press, Cambridge, UK, 1994.

Cassou, C., L. Terray and A.S. Phillips. Tropical Atlantic influence on European heat waves, *J. Climate*, 18, 2805–2811, 2005.

Cavenave, A. and R.S. Nerem. Present-day sea level change: Observations and causes, *Rev. of Geophy.*, 42, RG000139, 2004.

Chapin, F.S., M. Sturm, J.C. Serreze, J.P. McFadden, J.R. Key, A.H. Lloyd, A.D. McGuire, T.S. Rupp, A.H. Lynch, J.P. Schimel, J. Beringer, W.L. Chapman, J.E. Epstein, E.S. Euskirchen, L.D. Hinzman, J. Jia, C.-L Ping, K.D, Tape, C.D.C Thompson, D.A. Walker and J.M. Walker. Role of land-surface changes in Arctic summer warming, *Science*, 310, 657–660, Oct. 28, 2005.

Chen, J.L., C.R. Wilson and B.D. Tapley. Satellite gravity measurements confirm accelerated melting of Greenland Ice Sheet, *Science*, 313, 1958–1960, 29 Sept. 2006.

Chinn, T.J. New Zealand glacier responses to climate change of the past century, *New Zealand J. Geol. Geophy.*, 39, 415–428, 1996.

Christensen, T.R., T. Johansson, J.J. Akerman, M. Masepanov, N. Malmer, T. Friborg, P. Crill and B.H. Svensson. Thawing sub-arctic permafrost: Effects on vegetation and methane emissions, *Geophys. Res. Letters*, 31, L04501, 2004.

Christidis, N., P.A. Stott, S. Brown, G.C. Hegerl and J. Caesar. Detection of changes in temperature extremes during the second half of the 20th century, *Geophys. Res. Letters*, 32, L20716, 2005.

Christoffersen, P. and M.J. Hambrey. Is the Greenland ice sheet in a state of collapse?, *Geology Today*, 22, No. 3, 98–103, May–Jun. 2006.

Church, J.A. and N.J. White. A 20th century acceleration in global sea-level rise, *Geophys. Res. Letters*, 33, L01602, 2006.

Ciais, Ph. and 32 co-authors. Europe-wide reduction in primary productivity caused by the heat and drought in 2003, *Nature*, 437, Sep. 22, 2005.

Comiso, J.C. A rapidly declining perennial sea ice cover in the Arctic, *Geophys. Res. Letters*, 29, 17-1 to 17-4, 2002.

Comiso, J.C. Abrupt decline in the Arctic winter sea ice cover, *Geophys, Res. Letters*, 33, L18504, 2006.

Cook, E.R., L.J. Graumlich, P. Martin, J. Pastor, I.C. Prentice, T.R. Swetnam, K. Valentin, M. Verstraete, T. Webb III, J.White and I. Woodward. Biosphere-climate interactions during the past 18,000 years: Towards a global model of the terrestrial biosphere, in *Global Changes of the Past*, ed. R.S. Bradley, UCAR/OIES, 25–42, 1991.

Covey, C. The Earth's orbit and the ice ages, *Sci. Amer.*, 250, 58–66, 1984.

Crabtree, G.W., M.S. Dresselhaus and M.V. Buchanan. The hydrogen economy, *Physics Today*, 57, 39–44, Dec. 2004.

Crowley, T.J. Causes of climate change over the past 1000 years, *Science*, 289, 270–277, Jul. 14, 2000.

Curry, R. and C. Mauritzen. Dilution of the northern North Atlantic Ocean in recent decades, *Science*, 308, 1772–1774, Jun. 17, 2005.

Davis, C.H., Y. Li, J.R. McConnell, M.M. Frey and E. Hanna. Snowfall-driven growth in East Antarctic ice sheet mitigates recent sea-level rise, *Science*, 308, 1898–1901, Jun. 2005.

De Angelis, H. and P. Skvarca. Glacier surge after ice shelf collapse, *Science*, 1560–1562, Mar. 7, 2003.

Delworth, T.L., V. Ramaswamy and G.L. Stenchikov. The impact of aerosols on simulated ocean temperature and heat content in the 20th century, *Geophys. Res. Letters*, 32, L24709, 2005.

Dethloff, K and 14 others. A dynamical link between the Arctic and the global climate system, *Geophys. Res. Letters*, 33, L03703, 2006.

Dickson, B., I. Yashayaev, B. Turrell, S. Dye and J. Holfort. Rapid freshening of the deep North Atlantic Ocean over the past four decades, *Nature*, 416, 832, Apr. 25, 2002.

Dlugokencky, E.J., S. Houweling, L. Bruhwiler, K.A. Masarie, P.M. Lang, J.B. Miller and P.P. Tans. Atmospheric methane levels off: Temporary pause or a new steady-state?, *Geophys. Res. Letters*, 30, ASC 5-1 to 5-4, 2003.

Domack, E., D. Duran, A. Leventer, S. Ishman, S. Doane, S. McCallum, D. Amblas, J. Ring, R. Gilbert and M. Prentice. Stability of the Larsen B ice shelf on the Antarctic Peninsula during the Holocene epoch, *Nature*, 436, Aug. 2, 2005.

Donnadieu, Y., G. Ramstein, F. Fluteau and J. Besse. Is high obliquity a plausible cause for Neoproterozoic glaciations? *Geophys. Res. Letters*, 29, 2127, 2002.

Dupont, T.K. and R.B. Alley. Assessment of the importance of ice-shelf buttressing to ice-sheet flow, *Geophys. Res. Letters*, 32, L04503, 2005.

Ehrenman, G. Current from currents, *Mech. Engineering*, 40–41, Feb. 2003.

Ekstrom, G., M. Nettles and V.C. Tsai. Seasonality and increasing frequency of Greenland glacial earthquakes, *Science*, 311, 1756–1758, Mar. 24, 2006.

Emanuel, K. Increasing destructiveness of tropical cyclones over the past 30 years, *Nature*, 436, 686–688, 2005.

Fagan, B. *The Long Summer: How Climate Changed Civilization*, Basic Books, New York, 2004.

Fitzharris, B.B. and 11 co-authors. The cryosphere changes and their impacts, in *Climate Change 1995; Impacts, Adaptation and Mitigation of Climate Change*, R.T. Watson, M.C. Zinhyowere and R.J. Moss (eds.), 1996.

Foukal. P., G. North and T. Wigley. A stellar view on solar variations and climate, *Science*, 306, 68–69, Oct. 1, 2004.

Friborg, T., H. Soegaard, T.R. Christensen, C.R. Lloyd and N.S. Panikov. Siberian wetlands: Where a sink is a source, *Geophys. Res. Letters*, 30, CLM 5-1 to 5-4, 2003.

Friedmann, S. J. Storing carbon in Earth, *Geotimes*, 16–20, Mar. 2003.

Fu, Q., C.M. Johanson, S.G. Warren and D.J. Seldel. Contribution of stratospheric cooling to satellite-inferred tropospheric temperature trends, *Nature*, 429, 55–58, May 6, 2004.

Fu, Q., C.M., Johanson, J.M. Wallace and T. Reichler. Enhanced mid-latitude tropospheric warming in satellite measurements, *Science*, 312, 1179, May 26, 2006.

Fyfe, J.C. Southern Ocean warming due to human influence, *Geophys. Res. Letters*, 33, L19701, 2006.

Garidel-Thoron, T., Y. Rosenthal, F. Bassinot and L. Beaufort. Stable sea surface temperatures in the western Pacific warm pool over the past 1.75 million years, *Nature*, 433, 294–298, Jan. 20, 2005.

Gedney, N., P.M. Cox and C. Huntingford. Climate feedback from wetland methane emissions, *Geophys. Res. Letters*, 31, L20503, 2004.

Gelbspan, R. *Boiling Point: How Politicians, Big Oil and Coal, Journalists and Activists Have Fueled the Climate Crisis — and What We Can Do to Avert Disaster*, Basic Books, New York, 2004.

Gent, P.R. Will the North Atlantic Ocean thermohaline circulation weaken during the 21st century?, *Geophys. Res. Letters*, 28, 1023–1026, 2001.

Giorgi, F. Climate change hot-spots, *Geophys. Res. Letters*, 33, L08707, 2006.

Goldenberg, S.B., C.W. Landsea, A.M. Mestas-Nunez and W.M. Gray. The recent increase in Atlantic hurricane activity: Causes and implications, *Science*, 293, 474–479, Jul. 20, 2001.

Goldman, E. Even in the high Arctic nothing is permanent, *Science* 297, 1493-1494, 2002.

Gomes, R., Levison, H.F., Tsiganis, K. and Morbidelli, A. Origin of the cataclysmic Late Heavy Bombardment of the terrestrial planets, *Nature*, 435, 466, May 26, 2005.

Goosse, H., V. Masson-Delmotte, H. Renssen, M. Delmotte, T. Fichefet, V. Morgan, T. van Ommen, D.K. Khim and B Stenni. A late medieval warm period in the Southern Ocean as a delayed response to external forcing?, *Geophys. Res. Letters*, 31, L06203, 2004.

Gregg, W.W., M.E. Conkright, P. Ginoux, J.E. O'Reilly and N.W. Casey. Ocean primary production and climate: Global decadal changes, *Geophys. Res. Letters*, 30, OCE 3-1 to 3-4, 2003.

Gregg, W.W., N.W. Casey and C.R. McClain. Recent trends in global ocean chlorophyll, *Geophys. Res. Letters*, 32, L03606, 2005.

Gregory, J.M., P. Huybrechtsa, A. Sarah and C.B. Rapera. Threatened loss of the Greenland ice-sheet. *Nature*, 423, 616, Apr. 3, 2004.

Gregory, J.M. and 17 co-authors. A model intercomparison of changes in the Atlantic thermohaline circulation in response to increasing atmospheric CO_2 concentration, *Geophys. Res. Letters*, 32, L12703, 2005.

Gurney, K.R. and 25 co-authors. Towards robust regional estimates of CO_2 sources and sinks using atmospheric transport models, *Nature*, 415, 626–230, Feb. 7, 2002.

Hansen, J. and M. Sato. Greenhouse gas growth rates, *Proc. Nat. Acad. Sci.*, 101, no. 46, 16109–16114, Nov. 16, 2004.

Hansen, J.E. A slippery slope: How much global warming constitutes "Dangerous anthropogenic interference"?, *Climate Change*, 68, no. 3, 269–279, Feb. 2005.

Hansen, J., L. Nazarenko, R. Ruedy, M. Sato, J. Willis, A. Del Genio, D. Koch, A. Lacis, K. Lo, S. Menon, T. Novakov, J. Periwitz, G. Russell, C.A. Schmidt and N. Tausnev. Earth's energy imbalance: Confirmation and implications, *Science*, 308, 1431–1435, Jun. 3, 2005.

Hare, B. *Potsdam Institute for Climate Impact Research Study*, Feb. 2005.

Heath, J., E. Ayres, M. Possell, R.D. Bardgett, H.I.J. Black, H. Grant, P. Ineson and G. Kerstiens. Rising atmospheric CO_2 reduces sequestration of root-derived soil carbon, Science, 309, 1711–1713, Sept. 9, 2005.

Hegerl, G.C. and N.L. Bindoff. Warming of the world's oceans, *Science*, 309, 254–255, Jul. 8, 2005.

Hegerl, G.C. T.J. Crowley, W.T. Hyde and D.J. Frame. Climate sensitivity constrained by temperature reconstructions over the past seven centuries, *Nature*, 440, 1029–1032, Apr. 20, 2006.

Heinz Center. *Evaluation of Erosion Hazards* (Leatherman, Chair), 205 pp., Washington, DC, Apr. 2000.

Hemming, S.R. Heinrich events: Massive late Pleistocene detritus layers of the North Atlantic and their global climate imprint, *Rev. Geophys.*, RG1005, 2003.

Hobbie, S.E., K.J. Nadelhoffer and P. Hogberg. A synthesis: The role of nutrients as constraints on carbon balances in boreal and Arctic regions, *Plant and Soil*, 242, 163–170, 2002.

Hoffman, P., Kaufman, A., G.P. Halverson and D.P. Schrag, A Neoproterozoic snowball Earth, *Science*, 281, 1342, 1998.

Hofmann, D.J., J.H. Butler, T.J. Conway, E.J. Dlugokenchy, J.W. Elkings, K. Masarie, S.A. Montzka, R.C. Schnell and P. Tans. Tracking climate forcing: The annual greenhouse gas index, *EOS*, 87, 46, 509-511, 14 Nov. 2006.

Houghton, R.A. and D.L. Skole. Annual fluxes of carbon from deforestation and regrowth in the Brazilian Amazon, *Nature*, 403, Jan. 20, 2000.

Howat, I.M., I. Joughin, S. Tulaczyk and S. Gogineni. Rapid retreat and acceleration of Helheim Glacier, east Greenland, *Geophys. Res. Letters*, 32, L024737, 2005.

Huang, S. 1851–2004 annual heat budget of the continental landmasses, *Geophys. Res. Letters*, 33, L04707, 2006.

Huey, R.B. and P.D. Ward. Hypoxia, global warming and terrestrial late Permian extinctions, *Science*, 308, 398–401, 2005.

Huybers, P. Comment on "Hockey sticks, principal components and spurious significance" by S. McIntyre and R. McKitrick, *Geophys. Res. Letters*, 32, L20705, 2005.

IPCC. *Climate Change 2001: The Scientific Basis*, Contribution of Working Group I to the Third Assessment Report of the Intergovernmental Panel on Climate Change [Houghton, J.T., Y. Ding, D.J. Griggs, M. Noguer, P.J. van der Linden, X. Dai, K. Maskell and C.A. Johnson (eds.)] Cambridge University Press, Cambridge, UK and New York, NY, 881 pp., 2001.

Intergovernmental Panel on Climate Change, 4th Assessment Report, United Nations, 2007.

International Climate Change Taskforce. *Meeting the Climate Challenge* (S. Byers and O.J. Snowe, Co-chairs), The Institute for Public Policy Research, London, England, Jan. 2005.

International Energy Agency, World Energy Outlook 2006, Paris, France, 2006.

Jacobson, M.Z., W.G. Colella and D.M. Golden. Cleaning the air and improving health with hydrogen fuel-cell vehicles, *Science*, 308, 1901–1905, Jun. 24, 2005.

Jenkyns, H.C., A. Forster, S. Schouten and J.S. Sinninghe Damste. High temperatures in the Late Cretaceous Arctic Ocean. *Nature*, 432, 888–892, Dec. 16, 2004.

Johannessen, O.M., K. Khvorostovsky, M.W. Miles and L.P. Bobylev. Recent ice-sheet growth in the interior of Greenland, *Science*, 310, 1013–1016, Nov. 11, 2005.

Jones, P.D. et al. Adjusting for sampling density in box grid land and ocean surface temperature time series, *J. Geophys. Res.*, 106, 3371–3380, 2001.

Jones, P.D., M. New, D.E. Parker, S. Martin and I.G. Rigor. Surface air temperature and its changes over the past 150 years, *Reviews of Geophys.*, 37, 173–199, 1999.

Jones, C.D., P.M. Cox, R. L. H. Essery, D.L. Roberts and M.J. Woodage. Strong carbon cycle feedbacks in a climate model with interactive CO_2 and sulphate aerosols, *Geophys. Res. Letters*, 30, GL016867, 2003.

Jones, G.S., S.F.B. Tett and P.A. Stott. Causes of atmospheric temperature change 1960–2000: A combined attribution analysis, *Geophys. Res. Letters*, 30, 32-1 to 32-4, 2003.

Jones, P.D. and M.E. Mann. Climate over the past millennia, *Reviews of Geophys.*, 42, Am. Geophys. Union, RG2002, 2004.

Jonzen, N. and 16 co-authors. Rapid advance of spring arrival in long-distance migratory birds, *Science*, 312, 1959–1961, Jun. 30, 2006.

Joos, F., G-K. Plattner, T.F. Stocker, A. Kortzinger and D.R. Wallace. Trends in marine dissolved oxygen: Implications for ocean circulation changes and the carbon budget, *EOS*, 84, 197, May 27, 2003.

Jorgenson, M.T., Y.L. Shur and E.R. Pullman. Abrupt increase in permafrost degradation in Arctic Alaska, *Geophys. Res. Letters*, 33, L02503, 2006.

Joughin, I., E. Rignot, C.E. Rosanova, B.K. Lucchitta and J. Bohlander. Timing of recent accelerations of Pine Island glacier, Antarctica, *Geophys. Res. Letters*, 30, GL017609, 2003.

Joughin, I., W. Abdalati and M. Fahnestock. Large fluctuations in speed on Greenland's Jakobshavn glacier, *Nature*, 432, 608–610, Dec. 2, 2004.

Kargel, J.S., M.J. Abrams, M.P. Bishop, A. Bush, G. Hamilton, H. Jiskoot, A. Kaab, H.H. Kieffer, E.M. Lee, F. Paul, F. Rau, B. Raup, J.F. Shroder, D. Soltesz, D. Stainforth, L. Stearns and R. Wessels. Multispectral imaging contributions to global land ice measurements from space, *Remote Sensing of Environment*, 99, no. 1–2, 187–219, 2005.

Karoly, D.J., K. Braganza, P.A. Stott, J.M. Arblaster, G.A. Meehl, A.J. Broccoli and K.W. Dixon. Detection of a human influence on North American climate, *Science*, 302, 1200–1203, Nov. 14, 2003.

Kaser, G., J. G Cogley, M. B. Dyurgerov, M. F. Meier, A. Ohmura. Mass balance of glaciers and ice caps: Consensus estimates for 1961–2004, *Geophys. Res. Letters*, 33, L19501, 4 Oct. 2006.

Kaspar, F., N. Kuhl, U. Cubasch and T. Litt. A model-data comparison of European temperatures in the Eemian interglacial, *Geophys. Res. Letters*, 32, GL022456, 2005.

Katz, M.E., D.K. Pak, G.R. Dickens and K.G. Miller. The source and fate of massive carbon input during the latest Paleocene thermal maximum, *Science*, 286, 1531–1533, Nov. 19, 1999.

Kaufman, Y.J., O. Boucher, D. Tanre, M. Chin, L.A. Remer and T. Takemura. Aerosols anthropogenic component estimated from satellite data, *Geophys. Res. Letters*, 32, L17804, 2005.

Keeling, R.F. and H.E. Garcia, The change in oceanic O_2 inventory associated with recent global warming, *Proc. Nat. Acad. Sci.*, 99, 7848–7853, 2002.

Kennett, J.P., K.G. Cannariato, I.L. Hendy and R.J. Behl. *Methane Hydrates in Quaternary Climate Change: The Clathrate Gun Hypothesis*, American Geophysical Union Books Board, Washington, DC, 2003.

Keppler, F., J.T.G. Hamilton, M. Brass and T. Rockmann. Methane emissions from terrestrial plants under aerobic conditions, *Nature*, 439, 187–191, Jan. 12, 2006.

Kiehl, J.T. and K.E. Trenberth. Earth's annual global mean energy budget, *Bull. Amer. Meteor. Soc.*, 78, 12, 1997.

Kienast, M., T.J.J Hanebuth, C. Pelejero and S. Steinke. Synchroneity of meltwater pulse 1a and the Bolling Warming: New evidence from the South China Sea, *Geology*, 31, 67–70, 2003.

King, J.C. and J.C. Comiso. The spatial coherence of interannual temperature variations in the Antarctic Peninsula, *Geophys. Res. Letters*, 30, L015580, 2003.

Kirschvink, J.L. Late Proterozoic low-latitude global glaciation: the Snowball Earth, in *The Proterozoic Biosphere*, edited by J.W. Schoff and C. Klein, pp. 51–52, Cambridge University Press, New York, 1992.

Kitoh, A. and S. Murakami. Tropical Pacific climate at the mid-Holocene and the Last Glacial Maximum simulated by a coupled ocean-atmosphere general circulation model, *Paleoceanography*, 17, no. 3, 1047, 2002.

Khromova, T.E., M.B. Dyurgerov and R.G. Barry. Late-twentieth century changes in glacier extent in the Ak-shirak Range, Central Asia, determined from historical data and ASTER imagery, *Geophys. Res. Letters*, 30, HLS 2-1 to 2-4, 2003.

Knorr, W., I.C. Prentice, J.I. House and E.A. Holland. Long-term sensitivity of soil carbon turnover to warming, *Nature*, 433, 298–301, Jan. 20, 2005.

Knutti, R., F. Joos, S.A. Muller, G. Plattner and T.F. Stocker. Probabilistic climate change projections for CO_2 stabilization profiles, *Geophys. Res. Letters*, 32, L20707, 2005.

Krabill, W., E. Hanna, P. Huybrechts, W. Abdalati, J. Cappelen, B. Csatho, E. Frederick, S. Manizade, C. Martin, J. Sonntag, R. Swift, R.

Thomas and J. Yungel. Greenland ice sheet: Increased coastal thinning, *Geophys. Res. Letters*, 31, L24402, 2004.

Kunkel, K.E., D.R. Easterling, K. Hubbard and K. Redmond. Temporal variations in frost-free season in the United States: 1895–2000, *Geophys. Res. Letters*, 31, L03201, 2004.

Kvenvolden, K.A. Potential effects of gas hydrate on human welfare, *Proc. Nat. Acad. Sci.*, 96, no. 7, 3420–3426, 1999.

Lau, W.K.M. and K-M Kim. How nature foiled the 2006 hurricane forecasts, *EOS*, 88, 9, 105–107, 27 Feb. 2007.

Lawrence, D.M. and A.G. Slater. A projection of severe near-surface permafrost degradation during the 21st century, *Geophys. Res. Letters*, 32, L24401, 2005.

Lean. J., J. Beer and R. Bradley. Reconstruction of solar irradiance since 1610: Implications for climate change, *Geophys. Res. Letters*, 22, 3195–3198, 1995.

Leatherman, S.P., B.C. Douglas and J.L. LaBrecque. Sea level and coastal erosion require large-scale monitoring, *EOS, Trans. Amer. Geophys. Union*, 84, 13, 2003.

Lefevre, N., A.J. Watson, A. Olsen, A.F. Rios, F.F. Perez and T. Johannessen. A decrease in the sink for atmospheric CO_2 in the North Atlantic, *Geophys. Res. Letters*, L07306, 2004.

LeGrande, A.N., G.A. Schmidt, D.T. Shindell, C.V. Field, R.L. Miller, D.M. Koch, G. Faluvegi and G. Hoffmann. Consistent simulations of multiple proxy responses to an abrupt climate change event, *Proc. Nat. Acad. Sci.*, 10.1073/pnas. 0510095103, Jan. 13, 2006.

Leifer, I., B.P. Luyendyk, J. Boles and J.E. Clark. Natural marine seepage blowout: Contribution to atmospheric methane, *Global Biogeochem. Cycles*, 20, no. 3, GB308, July 20, 2006.

Lenton, T.M. and A.J. Watson. Biotic enhancement of weathering, atmospheric oxygen and carbon dioxide in the Neoproterozoic, *Geophys. Res. Letters*, 31, L05202, 2004.

Levitus, S., J.I. Antonov, J. Wang, T.L. Delworth, K.W. Dixon and A.J. Broccoli. Anthropogenic warming of Earth's climate system, *Science*, 292, 267–270, 2001.

Levitus, S., J.I. Antonov and T.P. Boyer. Warming of the world ocean, 1955–2003, *Geophys. Res. Letters*, 32, L02604, 2005.

Linden, E. *The Winds of Change*, Simon & Schuster, New York, NY, 2006.

Lindsay, R.W. and J. Zhang. The thinning of Arctic sea ice, 1988–2003: Have we passed a tipping point?, *J. Climate*, 18, 4879–4894, 2005.

Liu, J. and J. Diamond. China's environment in a globalizing world, *Nature*, 435, 1179–1186, Jun. 30, 2005.

Lui, Y. North Pacific warming and intense northwestern U.S. wildfires, *Geophys. Res. Letters*, 33, L21710, 2006.

Long, S.P., E.A. Ainsworth, A.D.B. Leakey, J. Nosberger and D.R. Ort. Food for thought: Lower-than-expected crop yield stimulation with rising CO_2 concentrations, *Science*, 312, 1918–1921, Jun. 30, 2006.

Luckman, A., T. Murray, R. de Lange and E. Hanna. Rapid and synchronous ice-dynamic changes in East Greenland, *Geophys. Res. Letters*, 33, L03503, 2006.

Luterhacher, J., D. Dietrich, E. Xoplaki, M. Grosjean and H. Wanner. European seasonal and annual temperature variability, trends and extremes since 1500, *Science*, 303, Mar. 5, 2004.

Mack, J.C., E.A.G. Schuur, M.S. Bret-Harte, G.R. Shaver and F.S. Chapin III. Ecosystem carbon storage in arctic tundra reduced by long-term nutrient fertilization, *Nature*, 431, 440–443, Sep. 23, 2004.

MacKenzie, F.T. *Our Changing Planet*, Prentice Hall, New Jersey, 1998.

Mann, M.E., R.S. Bradley and M.K. Hughes. Global-scale temperature patterns and climate forcing over the past six centuries, *Nature*, 392, 779–787, Apr. 23, 1998.

Mann, M.E., R.S. Bradley and M.K. Hughes. Northern hemisphere temperatures during the past millennium: Inferences, uncertainties and limitations, *Geophys. Res. Letters*, 26, 759–762, 1999.

Mann, M.E and P.D. Jones. Global surface temperatures over the past two millennia, *Geophys. Res. Letters*, 30, CLM 5–1 to 5–4, 2003.

Mann, M.E. and K.A. Emanuel. Atlantic hurricane trends linked to climate change, *EOS*, 87, no. 24, 233, Jun. 13, 2006.

Marchant, D.R., A.R. Lewis, W.M. Phillips, E.J. Moore, R.A. Souchez, G.H. Denton, D.E. Sugden, J. Potter Jr and G.P. Landis. Formation of patterned ground and sublimation till over Miocene glacier ice in Beacon Valley, southern Victoria Land, Antarctica, *GSA Bulletin*, 114, 718–730, 2002.

Martens, P. et al. Potential impacts of climate change on malaria risk, *Environmental Health Perspective*, 103 (5), 458–464, 1995.

Mastrandrea, M.D. and S.H. Schneider. Probabilistic integrated assessment of "Dangerous" climate change, *Science*, 304, 571–575, Apr. 23, 2004.

McCabe, G., M.A. Palecki and J.L. Betancourt. Pacific and Atlantic Ocean influences on multidecadal drought frequency in the United States, *Proc. Nat. Acad. Sci.*, 101, 4136–4141, 2004.

McIntyre, S. and McKitrick. Hockey sticks, principal components and spurious significance, *Geophys. Res. Letters*, 32, L03710, 2005.

McManus, J.F., R. Francois, J.-M. Gherardi, L.D. Keigwin and S.

Brown-Leger. Collapse and rapid resumption of Atlantic meridional circulation linked to deglacial climate changes, *Nature*, 428, 834–837, Apr. 22, 2004.

McMichael, A.J. et al. *Climate Change and Human Health*—Risks and Responses, UN World Health Organization, 2003.

McNeil, B.I., R.J. Matear and D.J. Barnes. Coral reef calcification and climate change: The effect of ocean warming, *Geophys. Res. Letters*, 31, L22309, 2004.

Meehl, G.A. and C. Tebaldi. More intense, more frequent and longer lasting heat waves in the 21st century, *Science*, 305, 994–997, Aug. 13, 2004.

Meehl, G.A. et al. How much more global warming and sea level rise?, *Science*, 307, 1769–1772, Mar. 18, 2005.

Meier, M.F. and M.B. Dyurgerov. How Alaska affects the world, *Science*, 297, 350–351, Jul. 19, 2002.

Meier, W., J. Stroeve, F. Fetterer and K. Knowles. Reductions in Arctic sea ice cover no longer limited to summer, *EOS, Trans. Amer. Geophys. Union*, 86, 326, 2005.

Meredith, M.P. and J.C. King. Rapid climate change in the ocean west of the Antarctic Peninsula during the second half of the 20th century, *Geophys. Res. Letters*, 32, L19604, 2005.

Millennium Ecosystem Assessment, Island Press, Washington, DC, 800 pp., Jan. 2006.

Milly, P.C.D., K.A. Dunne and A.V. Vecchia. Global pattern of trends in streamflow and water availability in a changing climate, *Nature*, 438, 347–350, Nov. 2005.

Moberg, A., D.M. Sonechkin, K. Holmgren, N.A. Datsenko and W. Karlen. Highly variable Northern Hemisphere temperatures reconstructed from low- and high-resolution proxy data, *Nature*, 433, 613–617, 10 Feb. 2005.

Moran, K. and 36 co-authors. The Cenozoic palaeoenvironment of the Arctic Ocean, *Nature*, 441, 601–605, Jun. 1, 2006.

Mote, P.W. Trends in snow water equivalent in the Pacific northwest and their climate causes, *Geophys. Res. Letters*, 30, GL017258, 2003.

Mouillot, F., A. Narasimha, Y. Balkanski, J.-F. Lamarque and C.B. Field. Global carbon emissions from biomass burning in the 20th century, *Geophys. Res. Letters*, 33, L01801, 2006.

Myers, R.A. and B. Worm. Rapid worldwide depletion of predatory fish communities, *Nature*, 423, 280–283, May 15, 2003.

National Research Council (US), Committee on Abrupt Climate Change (R. Alley, Chair). *Abrupt Climate Change: Inevitable Surprises*, National

Academy of Sciences, National Academy Press, Washington, DC, 2002.

National Research Council (US), Committee on Surface Temperature Reconstructions (R. North, Chair), *Surface Temperature Reconstructions for the Last 2,000 Years*, National Academy Press, Washington, DC, 2006.

Norris, R.D. and U. Röhl. Carbon cycling and chronology of climate warming during the Palaeocene/Eocene transition, *Nature*, 401, 775–778, Oct. 21, 1999.

North Greenland Ice Core Project members. High-resolution record of Northern Hemisphere climate extending into the last interglacial period, *Nature*, 431, 147–151, Sept. 9, 2004.

Oerlemans, J. Extracting a climate signal from 169 glacier records, *Science*, 308, 675–677, Apr. 29, 2005.

O'Neill, B.C. and M. Oppenheimer. Dangerous climate impacts and the Kyoto Protocol, *Science*, 296, 36, Jun. 14, 2002.

Oram, D.E., W.T. Sturges, S.A. Penkett, A. McCulloch and P.J. Fraser. Growth of fluoroform (CHF_3, HFC-23) in the background atmosphere, *Geophys. Res. Letters*, 25, 35–38, 1998.

Oreskes, N. The scientific consensus on climate change, *Science*, 306, 1686, Mar. 12, 2004.

Orr, A., D. Cresswell, G.J. Marshall, J.C.R. Hunt, J. Sommeria, C.G. Wang and M. Light. A "low-level" explanation for the recent large warming trend over the western Antarctic Peninsula involving blocked winds and changes in zonal circulation, *Geophys. Res. Letters*, 31, L06204, 2004.

Orr, J.C. and 26 co-authors. Anthropogenic ocean acidification over the twenty-first century and its impact on calcifying organisms, *Nature*, 437, Sep. 29, 2005.

Osborn, T.J. and K.R. Briffa. The spatial extent of 20th-century warmth in the context of the past 1200 years, *Science*, 311, 841–844, Feb. 10, 2006.

Otto-Bliesner, B.L., S.J. Marshall, J.T. Overpeck, G.H. Miller, A. Hu and CAPE Last Interglacial Project members. Simulating Arctic climate warmth and ice-field retreat in the last interglaciation, *Science*, 311, 1751–1753, Mar. 24, 2006.

Overpeck, J.T. and 20 co-authors. Arctic system on trajectory to new, seasonally ice-free state, *EOS*, 86, no. 34, Aug. 23, 2005.

Overpeck, J.T., B.L. Otto-Bliesner, G.H. Miller, D.R. Muhs, R.B. Alley and J.T. Kiehl. Paleoclimatic evidence for future ice-sheet instability and rapid sea-level rise, *Science*, 311, 1747–1750, Mar. 24, 2006.

Page, S.E., F. Siegert, J.O. Rieley, H.-D.V. Boehm, A. Jaya and S. Limin. The amount of carbon released from peat and forest fires in Indonesia during 1997, *Nature*, 420, 61–65, Nov. 7, 2002.

Parekh, P., S. Butkiewicz, M.J. Follows and T. Ito. Atmospheric carbon dioxide in a less dusty world, *Geophys. Res. Letters*, 33, L03610, 2006.

Parmesan, C. and G. Yohe. A globally coherent fingerprint of climate change impacts across natural systems, *Nature*, 421, 37–42, Jan. 2, 2003.

Paul, F., A. Kaab, M. Maisch, T. Kellenberger and W. Haeberli. Rapid disintegration of Alpine glaciers observed with satellite data, *Geophys. Res. Letters*, 31, L21402, 2004.

Patz, J.A., D. Campbell-Lendrum, T. Holloway and J.A. Foley. Impact of regional climate change on human health, *Nature*, 438, 310–317, Nov. 17, 2005.

Payne, A.J., A. Vieli, A.P. Shepherd, D.J. Wingham and E. Rignot. Recent dramatic thinning of largest West Antarctic ice stream triggered by ocean, *Geophys. Res. Letters*, 31, L23401, 2004.

Pearce, F. Climate warming as Siberia melts, *New Scientist*, Aug. 11, 2005.

Perkins, S. Once upon a lake: The life, times, and demise of the world's largest lake, *Science*, 162, 283–284, Nov. 2, 2002.

Perry, A.L., P.J. Low, J.R. Ellis and J.D. Reynolds. Climate change and distribution shifts in marine fishes, *Science*, 308, 1912–1915, 2005.

Petit, J.R. et al. Climate and atmospheric history of the past 420,000 years from the Vostok ice core, *Nature*, 399, 429–436, 1999.

Pezza, A.B. and I. Simmonds. The first South Atlantic hurricane: Unprecedented blocking, low shear and climate change, *Geophys. Res. Letters*, 32, L15712, 2005.

Philipona, R., B. Durr, C. Marty, A. Ohmura and M. Wild. Radiative forcing—measured at Earth's surface—corroborate the increasing greenhouse effect, *Geophys. Res. Letters*, 31, L03202, 2004.

Plattner, G.-K., F. Joos and T.F. Stocker. Revision of the global carbon budget due to changing air–sea oxygen fluxes, *Global Biogeochem. Cycles*, 16, BB001746, 2002.

Polyakov, I.V. and 22 co-authors. One more step toward a warmer Arctic, *Geophys. Res. Letters*, 32, GL023740, 2005.

Poulsen, C.J. A balmy Arctic, *Nature*, 432, 814–815, Dec. 16, 2004.

Pounds, J.A. and 13 co-authors. Widespread amphibian extinctions from epidemic disease driven by global warming, *Nature*, 439, 161–167, Jan. 12, 2006.

Prabhakara, C., R. Iacovazzi, J.-M. Yoo and G. Dalu. Global warming: Evidence from satellite observations, *Geophys. Res. Letters*, 27, 3517–3520, 2000.

Rahmstorf, S. The thermohaline ocean circulation: A system with dangerous thresholds? *Climate Change*, 46, 247–256, 2000.

Rahmstorf, S. A semi-empirical approach to projecting future sea-level rise, *Science*, 315, 368-370, 19 Jan. 2007.

Reichert, B.K., R. Schnur and L. Bengtsson. Global ocean warming tied to anthropogenic forcing, *Geophys. Res. Letters*, 29, 20-1 to 20-4, 2002.

Rignot, E., G. Casassa, P. Gogineni, W. Krabill, A. Rivera and R. Thomas. Accelerated ice discharge from the Antarctic Peninsula following the collapse of Larsen B ice shelf, *Geophys. Res. Letters*, 31, L18401, 2004.

Rignot, E., G. Casassa, S. Gogineni, P. Danagaratnam, W. Drabill, H. Pritchard, A. Rivera, R. Thomas, J. Turner and D. Vaughan. Recent ice loss from the Fleming and other glaciers, Wordie Bay, West Antarctic Peninsula, *Geophys. Res. Letters*, 32, L07502, 2005.

Rignot, E. and P. Kanagaratnam. Changes in the velocity structure of the Greenland ice sheet, *Science*, 311, 986–990, Feb. 17, 2006.

Rind, D., J. Lean and R. Healy. Simulated time-dependent climate response to solar radiative forcing since 1600, *J. Geophys. Res.*, 104, 1973–1990, Jan. 27, 1999.

Robeson, S.M. Trends in time-varying percentiles of daily minimum and maximum temperature over North America, *Geophys. Res. Letters*, 31, L04203, 2004.

Root, T.L., J.T. Price, K.R. Hall, S.J. Schneider, C. Rosenzweig and J.A. Pounds. Fingerprints of global warming on wild animals and plants, *Nature*, 421, 57–60, Jan. 2, 2003.

Rosenzweig, C., M.L. Parry, G. Fischer and K. Frohberg. *Climate Change and World Food Supply*. Research Report No. 3. Oxford: University of Oxford, Environmental Change Unit, 1993.

Rotstayn, L.D. and Y. Liu. A smaller global estimate of the second indirect aerosol effect, *Geophys. Res. Letters*, 32, L05708, 2005.

Rowley, R.J., J.C. Kostelnick, D. Braaten, X. Li, and J. Meisel. Risk of rising sea level to population and land area, *EOS*, 88, 9, 27 Feb. 2007.

Ruddiman, W.F. *Earth's Climate: Past and Future*, W.H. Freeman Press, 2001.

Ruddiman, W.F. and J.S. Thomson. The case for human causes of increased atmospheric CH_4 over the last 5000 years, *Quarterly Sci. Rev.*, 20, 1769–1777, 2001.

Sarmiento, J.L. and N. Gruber. Sinks for anthropogenic carbon, *Physics Today*, 30–36, Aug. 2002.

Sauber, J., B. Molnia, C. Carabajal, S. Lutheke and R. Muskett. Ice elevations and surface change on the Malaspina Glacier, Alaska, *Geophys. Res. Letters*, L23S01, 2005.

Scambos, T.A., J.A. Bohlander, C.A. Shuman and P. Skvarca. Glacier acceleration and thinning after ice shelf collapse in the Larsen B embayment, Antarctica, *Geophys. Res. Letters*, 31, L18402, 2004.

Scheffer, M., V. Brovkin and P.M. Cox. Positive feedback between global warming and atmospheric CO_2 concentration inferred from past climate change, *Geophys. Res. Letters*, 33, L10702, May 28, 2006.

Schellnhuber, H.J., W. Cramer, N. Nakicenovic, T. Wigley and G. Yohe (eds.). *Avoiding Dangerous Climate Change*, Cambridge University Press, 2006.

Schneider-Mor, A., R. Yam, C. Bianchi, M. Kunz-Pirrung, R. Gersonde and A. Shemesh. Diatom stable isotopes, sea ice presence and sea surface temperature records of the past 640 ka in the Atlantic sector of the Southern Ocean, *Geophys. Res. Letters*, 32, L10704, 2005.

Schulze, E.D. and A. Freibauer. Carbon unlocked from soils, *Nature*, 437, 205–206, Sep. 8, 2005.

Scientific Expert Group Report on Climate Change and Sustainable Development, Confronting Climate Change: Avoiding the Unmanageable and Managing the Unavoidable, United Nations Foundation, Sigma Xi, 2007.

Scotese, C.R. *Paleogeographic Atlas*, University of Texas Press, Austin, TX, 1997.

Service, R.S. As the west goes dry, *Science*, 303, 1124–1127, 2004.

Shaver, G.R., W.D. Billings, F.S. Chapin III, A.D. Giblin, K.J. Nadelhoffer, W.C. Oechel and E.B. Rastetter. Global change and the carbon balance of Arctic ecosystems, *BioScience*, 42, 433–441, 1992.

Shepherd, A., D. Wingham, T. Payne and P. Skvarca. Larsen ice shelf has progressively thinned, *Science*, 302, 856–858, Oct. 31, 2003.

Shepherd, A., D. Wingham and E. Rignot. Warm ocean is eroding West Antarctic Ice Sheet, *Geophys. Res. Letters*, 31, L23402, 2004.

Shindell, D.T. and G.A. Schmidt. Southern Hemisphere climate response to ozone changes and greenhouse gas increases, *Geophys. Res. Letters*, 31, L18209, 2004.

Shindell, D.T., B.P. Walter and G. Faluvegi. Impacts of climate change on methane emissions from wetlands, *Geophys. Res. Letters*, 31, L21202, 2004.

Shindell, D.T., G. Faluvegi, N. Bell and G.A. Schmidt. An emissions-based view of climate forcing by methane and tropospheric ozone, *Geophys. Res. Letters*, 32, GL021900, 2005.

Shukla, J., T. DelSole, M. Fennessy, J. Kinter and D. Paoline. Clinate model fidelity and projections of climate change, *Geophy. Res. Letters*, 33, L07702, 2006.

Siegenthaler, U., T.F. Stocker, E. Monnin, D. Luthi, J. Schwander, B. Stauffer, D. Raynaud, J.-M. Barnola, J. Fischer, V. Masson-Delmotte and J. Jouzel. Stable carbon cycle-climate relationship during the late Pleistocene, *Science*, 310, 1313–1317, Nov. 25, 2005.

Skvarca, P., W. Rack, J. Rott and T. Ibarzabal y Donanglo. Climate trend, retreat and disintegration of ice shelves on the Antarctic Peninsula: An overview, *Polar Res.* 18, 151–157, 1999.

Sluijs, A. and 14 co-authors. Subtropical Arctic Ocean temperatures during the Palaeocene/Eocene thermal maximum, *Nature*, 441, 610–613, Jun. 1, 2006.

Smith, L.C., Y. Sheng, G.M. MacDonald and L.D. Hinzman. Disappearing Arctic lakes, *Science*, 308, Jun. 3, 2005.

Socolow, R.H. Can we bury global warming, *Sci. Amer.*, 49–55, July 2005.

Sowers, T. Late Quaternary atmospheric CH_4 isotope record suggests marine clathrates are stable, *Science*, 311, 838, Feb. 10, 2006.

Spahni, R., J. Chappellaz, T.F. Stocker, L. Loulergue, G. Hausammann, K. Kawamura, J. Fluckiger, J. Schwander, D. Raynaud, V. Masson-Delmotte and J. Jouzel. Atmospheric methane and nitrous oxide of the late Pleistocene from Antarctic ice cores, *Science*, 310, 1317–1321, Nov. 25, 2005.

Stainforth, D.A., T. Aina, C. Christensen, M. Collins, N. Faull, D.J. Grame, J.A. Kettleborough, S. Knight, A. Martin, J.M. Murphy, C. Piani, D. Sexton, L.A. Smith, R.A. Spicer, A.J. Thorpe and M.R. Allen. Uncertainty in predictions of the climate response to rising levels of greenhouse gases, *Nature*, 433, 403–406, Jan. 27, 2005.

Stanhill, G. and S. Cohen. Global dimming: A review of the evidence for a widespread and significant reduction in global radiation with discussion of its probable causes and possible agricultural consequences, *Agric. For. Meteorol.*, 107, 155–278, 2001.

Stendel, M. and J.H. Christensen. Impact of global warming on permafrost conditions in a coupled GCM, *Geophys. Res. Letters*, 29, 10-1 to 10-4, 2002.

Stern, N., *Stern Review: The Economics of Climate Change*, Cambridge University Press, Cambridge, UK, 2006.

Stott, P.A. Attribution of regional-scale temperature changes to anthropogenic and natural causes, *Geophys. Res. Letters*, 30, 2-1 to 2-4, 2003.

Stroeve, J.C., J.C. Serreze, F. Fretterer, T. Arbetter, W.N. Meier, J. Maslanik and K. Knowles. Tracking the Arctic's shrinking ice cover: Another extreme minimum in 2004, *Geophys. Res. Letters*, 32, L04501, 2005.

Strom, R., R. Malhotra, T. Ito, F. Yoshida and D. Kring. The origin of planetary impactors in the inner solar system, *Science*, 309, 1847, Sep. 16, 2005.

Suess, E., G. Bohrmann, J. Greinert and E. Lausch. Flammable ice, *Sci. Amer.*, 76–83, Nov. 1999.

Sutton, R.T. and Hodson, D.L.R. Atlantic Ocean forcing of North American and European summer climate, *Science*, 309, 115–118, Jul. 1, 2005.

Taylor, R.G., L. Mileham, C. Tindimugaya, A. Majugu, A. Muwanga and B. Nakileza. Recent glacial recession in the Rwenzori Mountains of East Africa due to rising air temperature, *Geophys. Res. Letters*, 33, L10402, 2006.

Thomas, J.A., M.G. Telfer, D.B. Roy, C.D. Preston, J.J.D. Greenwood, J. Asher, R. Fox, R.T. Clarke and J.H. Lawton. Comparative losses of British butterflies, birds and plants and the global extinction crisis, *Science*, 303, 1879–1881, Mar. 19, 2004.

Thomas, R.H., E. Rignot, P. Kanagaratnam, W. Krabill and G. Casassa. Force-perturbation analysis of Pine Island Glacier, Antarctica, suggests cause for recent acceleration, *Annals of Glaciology*, 39, 133–138, 2004.

Thomas, D.S.G., M. Knight and G.F.S. Wiggs. Remobilization of southern African desert dune systems by twenty-first century global warming, *Nature*, 435, Jun. 30, 2005.

Thomas, R., E. Frederick, W. Krabill, S. Manizade and C. Martin. Progressive increase in ice loss from Greenland, *Geophys. Res. Letters*, 33 L10503, May 28, 2006.

Torn, M.S. and J. Harte. Missing feedbacks, asymmetric uncertainties and the underestimation of future warming, *Geophys. Res. Letters*, 33, L10703, May 28, 2006.

Trenberth, K.E. (ed.). *Climate System Modelling*, Cambridge University Press, Cambridge, UK, 1992.

Trenberth, K.E. and D.J. Shea. Atlantic hurricanes and natural variability in 2005, *Geophys. Res. Letters*, 33, no. 12, L12704, Jun. 27, 2006.

Treydte, K.S., G.H. Schleser, G. Helle, D.C. Frank, J. Winiger, G.H. Haug and J. Esper. The twentieth century was the wettest period in northern Pakistan over the past millennium, *Nature*, 440, 1179–1182, Apr. 27, 2006.

Tripati, A. and H. Elderfield. Deep-sea temperature and circulation changes at the Paleocene–Eocene thermal maximum, *Science*, 308, 1894–1898, Jun. 24, 2005.

Trombulak, S.C. and R. Wolfson. Twentieth-century climate change in

New England and New York, USA, *Geophys. Res. Letters*, 31, L19202, 2004.

Turetsky, M., K. Wieder, L. Halsey and D. Vitt. Current disturbance and the diminishing peatland carbon sink, *Geophys. Res. Letters*, 29, 21-1 to 21–4, 2002.

Turner, J., S.R. Colwell, G.J. Marshall, T.A. Lachlan-Cope, A.M. Carleton, P.D. Jones, V. Lagun, P.A. Reid and S. Iagovkina. Antarctic climate change during the last 50 years, *Internat. J. Climatology*, 25, 279–294, 2005.

Turner, J., T.A. Lachian-Cope, S. Colwell, G.J. Marshall and W.M. Connolley. Significant warming of the Antarctic winter troposphere, *Science*, 311, 1914–1917, Mar. 31, 2006.

Tyedmers, P.H., R. Watson and D. Pauly. Fueling global fishing fleets, *Ambio*, 34, no. 8, 635–638, Dec. 2005.

Valley, J., W. Peck, E. King and S. Wilde. A cool early Earth, *Geology*, 30, 351, Apr. 2002.

Van Andel, T.H. *New Views on an Old Planet: A History of Global Change*, Cambridge University Press, Cambridge, UK, 1994.

Van den Broeke, M. Strong surface melting preceded collapse of Antarctic Peninsula ice shelf, *Geophys. Res. Letters*, 32, L12815, 2005.

Vanlooy, J., R. Forster and A. Ford. Accelerating thinning of Kenai Peninsula glacier, Alaska, *Geophys. Res. Letters*, 33, L21307. 2006.

Vargas-Yáñez, M., G. Parrilla, A. Lavin, P. Velez-Belchi and C. Gonzalez-Pola. Temperature and salinity increase in the eastern North Atlantic along the 24.5 °N in the last ten years, *Geophys. Res. Letters*, 31, L06210, 2004.

Vaughan, D.G. and C.S.M. Doake. Recent atmospheric warming and retreat of ice shelves on the Antarctic Peninsula, *Nature*, 379, 328–331, Jan. 1996.

Vecchi, G., B.J. Soden, A.T. Wittenberg, I.M. Held, A. Leetmaa and J.J. Harrison. Weakening of tropical Pacific atmospheric circulation due to anthropogenic forcing, *Nature*, 441, 73–76, May 4, 2006.

Velicogna, I. and J. Wahr. Greenland mass balance from GRACE, *Geophys. Res. Letters*, 32, L18505, 2005.

Velicogna, I. and J. Wahr. Measurements of time-variable gravity show mass loss in Antarctica, *Science*, 311, 1754–1756, Mar. 24, 2006.

Velicogna, I. and J. Wahr. Acceleration of Greenland ice mass loss in spring 2004, *Nature*, 443, 329-331, 21 Sept. 2006.

Vellinga, M. and R.A. Wood. Global climatic impacts of a collapse of the Atlantic thermohaline circulation, *Climate Change*, 54, no. 3, 251–267, 2002.

Walter, K.M., S.A. Zimov, J.P. Chanton, D. Verbyla and F.S. Chapin III. Methane bubbling from Siberian thaw lakes as a positive feedback to climate warming, *Nature*, 443, 71-75, 7 Sept. 2006

Webster, P.J., G.J. Holland, J.A. Curry and H.-R. Chang. Changes in tropical cyclone number, duration and intensity in a warming environment, *Science*, 309, 1844–1846, Sep. 16, 2005.

Wellington, G.M., P.W. Glynn, A.E. Strong, S.A. Navarrete, E. Wieters and D. Hubbard. Crisis on coral reefs linked to climate change, *EOS*, 82, no. 1, Jan. 2, 2001.

Weart, S. R. *The Discovery of Global Warming*, Harvard University Press, Cambridge, MA, 2003.

Westerling, A.L., H.G. Hidalgo, D.R. Cayan, T.W. Swetnam, Warming and earlier Spring increases Western US forest wildfire activity, *Science*, 312, Jul. 6, 2006.

Wetherald, R.T., R.J. Stouffer and K.W. Dixon. Committed warming and its implications for climate change, *Geophys. Res. Letters*, 28, 1535–1538, 2001.

White, N.J., J.A. Church and J.M. Gregory. Coastal and global averaged sea level for 1950 to 2000, *Geophys. Res. Letters*, 32, Jan. 2005.

Wiglely, T.M.L. The climate change commitment, *Science*, 307, 1766–1769, Mar. 18, 2005.

Winguth, A., U. Mikolajewicz, M. Groger, D. Maier-Reimer, G. Schurgersw and M. Vizcaino. Centennial-scale interactions between the carbon cycle and anthropogenic climate change using a dynamic Earth system model, *Geophys. Res. Letters*, L23714, 2005.

Woodgate, R.A. and K. Aagaard. Revising the Bering Strait freshwater flux into the Arctic Ocean, *Geophys. Res. Letters*, 32, L02602, 2005.

Worm, B. and 13 co-authors. Impacts of biodiversity loss on ocean ecosystem services, *Science*, 314, 787-790, 3 Nov. 2006.

Wu, P., R. Wood and P. Stott. Does the recent freshening trend in the North Atlantic indicate a weakening thermohaline circulation?, *Geophys. Res. Letters*, 31, L02301, 2004.

Young, J.J., S. Mehta, M. Israelsson, J. Dodoski, E. Grill and J.I. Schroeder. CO_2 signaling in guard cells: Calcium sensitivity response modulation, a Ca^{2+}-independent phase and CO_2 insensitivity of the *gca2* mutant, *Proc. Nat. Acad. Sci.*, 103, no. 19, 7506–7511, May 9, 2006.

Zachos, J.C., M.W. Wara, S. Bohaty, M.L. Delaney, M.R. Petrizzo, A. Brill, T.J. Bralower and I. Premoli-Silva. A transient rise in tropical sea surface temperature during the Paleocene–Eocene thermal maximum, *Science*, 302, 1551–1554, Nov. 28, 2003.

Zeebe, R.E. and D. Archer. Feasibility of ocean fertilization and its impact on future CO_2 levels, *Geophys. Res. Letters*, 32, L09703, 2005.

Zhang, Y. Methane excape from gas hydrate systems in marine environment and methane-driven oceanic eruptions, *Geophys. Res. Letters*, 30, 51-1 to 51-4, 2003.

Zhang. Y. and W. Chen. Soil temperature in Canada during the twentieth century: Complex responses to atmospheric climate change, *J. Geophys. Res.*, 110, D03112, 2005.

Zimov, S.A., E.A.G. Schuur and F.S. Chapin III. Permafrost and the global carbon budget, *Science*, 312, 1612–1613, Jun. 16, 2006.

Zimov, S.A, S.P. Davydov, G.M. Zimova, A.I. Davydova, E.A.G. Schuur, K. Dutta and F.S. Chapin III. Permafrost carbon: Stock and decomposability of a globally significant carbon pool, *Geophys. Res. Letters*, 33, L20502, 2006.

Zwally, H.J., W. Abdalati, T. Herring, K. Larson, J. Saba and K. Steffen. Surface melt-induced acceleration of Greenland ice-sheet flow, *Science*, 297, 218–226, Jul. 2002.

FURTHER READING

Abrupt Climate Changes

Bendtsen, J. and C.J. Bjerrum. Vulnerability of climate on Earth to sudden changes in insolation, *Geophys. Res. Letters*, 29, 1-1 to 1-4, 2002.

Dickens, G.R. Methane hydrate and abrupt climate change, *Geotimes*, 18–27, Nov. 2004.

Rahmstorf, S. Timing of abrupt climate change: A precise clock, *Geophys. Res. Letters*, 30, 17-1 to 17-4, 2003.

Rossby, T. and J. Nilsson. Current switching as the cause of rapid warming at the end of the last Glacial Maximum and Younger Dryas, *Geophys. Res. Letters*, 30, 23-1 to 23-4, 2003.

Schaeffer, M., F.M. Selten and J.D. Opsteegh. Intrinsic limits to predictability of abrupt regional climate change in IPCC SRES scenarios, *Geophys. Res. Letters*, 29, 14-1 to 14-4, 2002.

Aerosols

Schafer, J.S., T.F. Eck, B.N. Holben, P. Artaxo, M.A. Yamasoe and A.S. Procopio. Observed reductions of total solar irradiance by biomass-burning aerosols in the Brazilian Amazon and Zambian Savanna, *Geophys. Res. Letters*, 29, 4-1 to 4-4, 2002.

Venkataraman, C., G. Habib, A. Diguren-Fernandez, A.H. Miguel and S.K. Friedlander. Residential biofuels in South Asia: Carbonaceous aerosol emissions and climate impacts, *Science*, 307, 1454–1456, Mar. 4, 2005.

Ming, Y., L.M. Russel and D.F. Bradford. Health and climate policy impacts on sulfur emission control, *Rev. Geophys.*, 43, no. 4, RG4002, Dec. 2005.

Nagashima, T., H. Shiogama, T. Yokohata, T. Takemura, S.A. Crooks, and T. Nozawa. Effect of carbonaceous aerosols on surface temperature in the mid twentieth century, *Geophys. Res. Letters*, 33, L04702, 2006.

Biology

Board on Global Health, Institute of Medicine. *Microbial Threats to Health: Emergence, Detection, and Response*, The National Academies Press, 2003.

Boyd, P.W. and S.C. Doney, Modeling regional responses by marine pelagic ecosystems to global climate change, *Geophys. Res. Letters*, 29, 53-1 to 53-4, 2002.

Cook, E.R., L.J. Graumlich, P. Martin, J. Pastor, I.C. Prentice, T.R. Swetnam, K. Valentin, M. Verstraete, T. Webb III, J. White and I. Woodward. Biosphere–Climate interactions during the past 18,000 years: Towards a global model of the terrestrial biosphere, in *Global Changes of the Past*, ed. R.S. Bradley, UCAR/OIES, 25–42, 1991.

Epstein, P.R. Is global warming harmful to health?, *Sci. Amer.*, 50–57, Aug. 2000.

Gedney, N. and P.J. Valdes. The effect of Amazonian deforestation on the northern hemisphere circulation and climate, *Geophys. Res. Letters*, 27, 3053–3056, 2000.

Grebmeir, J.M., J.E. Overland, S.E. Moore, D.V. Farley, E.C. Carmack, L.W. Cooper, K.E. Frey, J.H. Helle, F.A. McLaughlin and S.L. McNutt. A major ecosystem shift in the northern Bering Sea, *Science*, 331, 1461–1464, Mar. 10, 2006.

Gregg, W.W. and M.E. Conkright. Decadal changes in global ocean chlorophyll, *Geophys. Res. Letters*, 29, 20-1 to 20-4, 2002.

Grossman, D. Spring forward, *Sci. Amer.*, 85–91, Jan. 2004.

Hoffmann, W.A., W. Schroeder and R.B. Jackson. Positive feedbacks of fire, climate, and vegetation and the conversion of tropical savanna, *Geophys. Res. Letters*, 29, 9-1 to 9-4, 2002.

Hughen, K.A., T.I. Eglinton, L. Xu and M. Makou. Abrupt tropical vegetation response to rapid climate change, *Science*, 304, 1955–1962, 2004.

Lewis, T.J. and K. Wang. Geothermal evidence for deforestation induced warming: Implications for the climate impact of land development, *Geophys. Res. Letters*, 25, 535–538, 1998.

Maslin, M. Ecological versus climate thresholds, *Science*, 306, 2197–2198, Dec. 24, 2004.

Matthews, H.D., A.J. Weaver, M. Eby and K.T. Meissner. Radiative forcing of climate by historical land cover change, *Geophys. Res. Letters*, 30, 27-1 to 27-4, 2003.

Mote, P.W. and N.J. Mantua. Coastal upwelling in a warmer future, *Geophys. Res. Letters*, 29, 53-1 to 53-4, 2002.

Neuer, S., R. Davenport, T. Freudenthal, G. Wefer, O. Llinas, M.-J. Rueda, D.K. Steinberg and D.M. Karl. Differences in the biological carbon pump at three subtropical ocean sites, *Geophys. Res. Letters*, 32-1 to 32-4, 2002.

Oyama, M.D. and C.A. Nobre. A new climate-vegetation equilibrium state for Tropical South America, *Geophys. Res. Letters*, 30, CLM 5-1 to 5-4, 2003.

Peterson, W.T. and F.B. Schwing. A new climate regime in northeast Pacific ecosystems, *Geophys. Res. Letters*, 30, OCE 6-1 to 6-4, 2003.

Piao, S., P. Friedlingstein, P. Ciais, L. Zhou and A. Chen. Effect of climate and CO_2 changes on the greening of the Northern Hemisphere over the past two decades, *Geophys. Res. Letters*, 33, L23402, 2006.

Ransom, A.M. and B. Worm. Rapid worldwide depletion of predatory fish communities, *Nature*, 423, 280–283, 2003.

Richardson, A.J. and D.S. Schoeman. Climate impact on plankton ecosystems in the Northeast Atlantic, *Science*, 305, 1609–1612, Sep. 10, 2004.

Thomas, C.D. and 18 other authors. Extinction risk from climate change, *Nature*, 427, 145–148, Jan. 8, 2004.

Thompson, S.L., B. Govindasamy, A. Mirin, K. Caldeira, C. Delire, J. Milovich, M. Wickett and D. Erickson. Quantifying the effects of CO_2-fertilized vegetation on future global climate and carbon dynamics, *Geophys. Res. Letters*, 31, L23211, 2004.

Timmermann, A. and Jin, F.-F. Phytoplankton influences on tropical climate, *Geophys. Res. Letters*, 29, 19-1 to 19-4, 2002.

Carbon Sinks

Baldocchi, D. The carbon cycle under stress, *Nature*, 437, Sep. 22, 2005.

Bates, N.R., S. B. Moran, D.A. Hansell and J.T. Mathis, An increasing CO_2 sink in the Arctic Ocean due to sea-ice loss, *Geophys. Res. Letters*, 33, L23609, 2006.

Davidson, E.A. and I.A. Janssens. Temperature sensitivity of soil carbon decomposition and feedbacks to climate change, *Nature*, 440, 165–173, Mar. 9, 2006.

Fallon, S.J., T.P. Guilderson and K. Caldeira. Carbon isotope constraints on vertical mixing and air–sea CO_2 exchange, *Geophys. Res. Letters*, 30, OCE 9-1 to 9-4, 2003.

Goulden, M.L., S.C. Wofsy, J.W. Harden, S.E. Trumbore, P.M. Crill, S.T. Gower, T. Fries, B.C. Daube, S.-M. Fan, D.J. Sutton, A. Bazzaz and J.W. Munger. Sensitivity of boreal forest carbon balance to soil thaw, *Science*, 279, 214–216, 1998.

Govindasamy, B., S. Thomson, A. Mirin, M. Wickett, K. Caldeira and C. Delire. Increase of carbon cycle feedback with climate sensitivity: results from a coupled climate and carbon cycle model, *Tellus*, 57B, 153–163, 2005.

Joos, F., G.-K. Plattner, T.F. Stocker, O. Marchal and A. Schmittner. Global warming and marine carbon cycle feedbacks on future atmospheric CO_2, *Science*, 284, 464–567, Apr. 16, 1999.

Joos, F., I.C. Prentice, S. Sitch, R. Meyer, G. Hooss, G.-K. Plattner, S. Gerber and K. Hasselmann. Global warming feedbacks on terrestrial carbon uptake under the Intergovernmental Panel on Climate Change (IPCC) emission scenarios, *Global Biogeochem. Cycles*, 15, no. 4, 891–907, Dec. 2001.

Kaplan, J.O., I.C. Prentice, W. Knorr and P.J. Valdes. Modeling the dynamics of terrestrial carbon storage since the last Glacial Maximum, *Geophys. Res. Letters*, 29, 31-1 to 31-4, 2002.

Kortzinger, A. A significant CO_2 sink in the tropical Atlantic Ocean associated with the Amazon River plume, *Geophys. Res. Letters*, 30, OCE 8-1 to 8-4, 2003.

Leahy, P., G. Kiely and T.M. Scanlon. Managed grasslands: A greenhouse gas sink or source?, *Geophys. Res. Letters*, 31, L20507, 2004.

Sabine, C.L. and 14 co-authors. The oceanic sink for anthropogenic CO_2. *Science*, 305, 367–371, Jul. 16, 2004.

Sarmiento, J.L. and N. Gruber. Sinks of anthropogenic carbon, *Physics Today*, 55, 30–36, Aug. 2002.

Schiermeier, Q. That sinking feeling, *Nature*, 435, 732–733, Jun. 9, 2005.

Wakita, M., Y.W. Watanabe, S. Watanabe and S. Noriki. Oceanic uptake rate of anthropogenic CO_2 in a subpolar marginal sea: The Sea of Okhotsk, *Geophys. Res. Letters*, 30, OCE 4-1 to 4-4, 2003.

Consequences

Benestad, R.E. Climate change scenarios for northern Europe from multi-model IPCC AR4 climate simulations, *Geophys. Res. Letters*, 32, L17704, 2005.

Chan, W.-L. and T. Motoi. Response of thermohaline circulation and thermal structure to removal of ice sheets and high atmospheric CO_2 concentration, *Geophys. Res. Letters*, 32, L07601, 2005.

Dai, A., T.M.L. Wigley, G.A. Meehl and W.M. Washington. Effects of stabilizing atmospheric CO_2 on global climate in the next two centuries, *Geophys. Res. Letters*, 28, 4511–4514, 2001.

European Environment Agency. *Impacts of Europe's Changing Climate*, European Topic Centre for Air and Climate Change, 2004.

Fowler, H.J. and C.G. Kilsby. Implications of changes in seasonal and annual extreme rainfall, *Geophys. Res. Letters*, 30, 53-1 to 53-4, 2003.

Haarsama, R.J., F.M. Selten, S.L. Weber and M. Klephuis. Sahel rainfall variability and response to greenhouse warming, *Geophys. Res. Letters*, 32, L17702, 2005.

Hayhoe, K. and 18 co-authors. Emissions pathways, climate change, and impacts on California, *National Academy of Sciences*, 101, no. 34, Aug. 24, 2004.

McHugh, M.J. Multi-model trends in East African rainfall associated with increased CO_2, *Geophys. Res. Letters*, 32, L01707, 2005.

National Assessment Synthesis Team. *Climate Change Impacts on the United States: The Potential Consequences of Climate Variability and Change*, U.S. Global Change Research Program, Cambridge University Press, 2000.

Pal, J.S., F. Giorgi and X. Bi. Consistency of recent European summer precipitation trends and extremes with future regional climate projections, *Geophys. Res. Letters*, 31, L13202, 2004.

Report of Working Group II. *Climate Change 2001: Impacts, Adaptation, and Vulnerability*, Intergovernmental Panel on Climate Change, 2001.

Snyder, M.A., J.L. Bell and L.C. Sloan. Climate responses to doubling of atmospheric carbon dioxide for a climatically vulnerable region, *Geophys. Res. Letters*, 20, 9-1 to 9-4, 2002.

Zolina, O., C. Simmer, A. Kapala and S. Gulev. On the robustness of the estimates of centennial-scale variability in heavy precipitation from station data over Europe, *Geophys. Res. Letters*, 32, L14707, 2005.

Country Emissions

Aldhous, P. China's burning ambition, *Nature*, 435, 1152–1154, Jun. 30, 2005.

Granier, C. and G.P. Brasseur. The impact of road traffic on global tropospheric ozone, *Geophys. Res. Letters*, 30, 58-1 to 58-4, 2003.

Millet, D.B. and A.H. Goldstein. Evidence of continuing methylchloroform emissions from the United States, *Geophys. Res. Letters*, 31, L17101, 2004.

Cryosphere (The Earth's Ice)

Bamber, J., W. Krabill, V. Raper and J. Dowdeswell. Anomalous recent growth of part of a large Arctic ice cap: Austfonna, Svalbard, *Geophys. Res. Letters*, 31, L12402. 2004.

Bindschadler, R.A., R.B. Alley, J. Anderson, S. Shipp, H. Borns, J. Fastook, S. Jacobs, C.F. Raymond and C.A. Shuman. What is happening to the West Antarctic ice sheet?, *EOS, Trans. Amer. Geophys. Union*, 79, 263–265, Jun. 1998.

Cavalieri, D.J., C.L. Parkinson and K.Y. Vinnikov. 30-Year satellite record reveals contrasting Arctic and Antarctic decadal sea ice variability, *Geophys. Res. Letters*, 30, CRY 4-1 to 4-4, 2003.

Clark, P.U., R.B. Alley and D. Pollard. Northern hemisphere ice-sheet influences on global climate change, *Science*, 286, 1104–1111, Nov. 5, 1999.

Comiso, J.C. A rapidly declining perennial sea ice cover in the Arctic, *Geophys. Res. Letters*, 29, 17-1 to 17-4, 2002.

Comiso, J.C. and C.L. Parkinson. Satellite-observed changes in the Arctic, *Physics Today*, 57, 38–44, Aug. 2004.

Corr, H.F.J., A. Jenkins, K.W. Nicholls and S.M. Doake. Precise measurements of changes in ice-shelf thickness by phase-sensitive radar to determine basal melt rates, *Geophys. Res. Letters*, 29, 73-1 to 73-4. 2002.

Cuffey, K.M., H. Conway, B. Hallet, A.M. Gades and C.F. Raymond. Interfacial water in polar glaciers and glacier sliding at 17 °C, *Geophys. Res. Letters*, 26, 751–754, 1999.

Curran, M.A.J., T.D. van Ommen, V.I. Morgan, K.L. Phillips and A.S. Palmer. Ice core evidence for Antarctic sea ice decline since the 1950s, *Science*, 302, 1203–1206, Nov. 14, 2003.

Dethloff, K., W. Dorn, A. Rinke, K. Fraedrich, M. Junge, E. Roechner, V. Gayler, U. Cubasch and J.H. Christensen. The impact of Greenland's deglaciation on the Arctic circulation, *Geophys. Res. Letters*, 31, L19201, 2004.

Favier, V., P. Wagnon and P. Ribstein. Glaciers of the outer and inner tropics: A different behavior but a common response to climate forcing, *Geophys. Res. Letters*, 31, L16403, 2004.

Fichefet, T., C. Poncin, H. Goosse, P. Huybrechts, I. Janssens and H. Le Treut. Implications of changes in freshwater flux from the Greenland ice sheet for the climate of the 21st century, *Geophys. Res. Letters*, 30, CLM 8-1 to 8-4, 2003.

Gray, L., I. Joughin, S. Tulaczyk, V.B. Spikes, R. Bindschadler and K. Jezek. Evidence for subglacial water transport in the West Antarctic Ice Sheet through three-dimensional satellite radar interferometry, *Geophys. Res. Letters*, 32, L021387, 2005.

Haas, C. Late-summer sea ice thickness variability in the Arctic Transpolar Drift 1991–2001 derived from ground-based electromagnetic sounding, *Geophys. Res. Letters*, 31, L09402, 2004.

Harper, J.T. and N.F. Humphrey. High altitude Himalayan climate from glacial ice flux, *Geophys. Res. Letters*, 30, HLS 3-1 to 3-4, 2003.

Hilmer, M. and P. Lemke. On the decrease of Arctic sea ice volume, *Geophys. Res. Letters*, 27, 3751–3754, 2000.

Holland, M.M., C.M. Bitz, and B. Tremblay. Future abrupt reductions in the summer Artic sea ice, *Geophys. Res. Letters*, 33, L23503, 2006.

Jia, G.J. and H.E. Epstein. Greening of arctic Alaska, 1981–2001, *Geophys. Res. Letters*, 30, 2067, 2003.

Larour, E., E. Rignot and D. Aubry. Modeling of rift propagation on Ronne Ice Shelf, Antarctica, and sensitivity to climate change, *Geophys. Res. Letters*, 31, L16404, 2004.

Marshall, S.J. and K.M. Cuffey. Peregrinations of the Greenland Ice Sheet divide in the last glacial cycle: Implications for central Greenland ice cores, *Earth Planetary Sci. Letters*, 179, 73–90, 2000.

Moore, G.W.K. Reduction in seasonal sea ice concentration surrounding southern Baffin Island 1979-2004, *Geophys. Res. Letters*, 33, L20501, 2006.

Mueller, D.R., W.F. Vincent and M.O. Jeffries. Break-up of the largest Arctic ice shelf and associated loss of an epishelf lake, *Geophys. Res. Letters*, 30, CRY 1-1 to 1-4, 2003.

Multiple authors. Polar science, *Science*, 297, 1491–1506, Aug. 30, 2002.

Muskett, R.R., C.S. Lingle, W.V. Tangborn and B.T. Rabus. Multi-decadal elevation changes on Bagley Ice Valley and Malaspina Glacier, Alaska, *Geophys. Res. Letters*, 30, CRY 1-1 to 1-4, 2003.

Oelke, C., T. Zhang and M.C. Serreze. Modeling evidence for recent warming of the Arctic soil thermal regime, *Geophys. Res. Letters*, 31, L07208, 2004.

Pederson, G.T., D.B. Fagre, S.T. Gray and L.J. Graumlich. Decadal-scale climate drivers for glacial dynamics in Glacier National Park, Montana, USA, *Geophys. Res. Letters*, 31, L12203, 2004.

Petit J.R., J. Jouzel, D. Raynaud, N.I. Barkov, J.M. Barnola, I. Basile, M. Bender, J. Chappellaz, J. Davis, G. Delaygue, M. Delmotte, V.M. Kotlyakov, M. Legrand, V. Lipenkov, C. Lorius, L. Pépin, C. Ritz, E. Saltzman and M. Stievenard. Climate and atmospheric history of the past 420,000 years from the Vostok Ice Core, Antarctica, *Nature*, 399, 429–436, 1999.

Price, P.B., K. Woschnagg and D. Chirkin. Age *vs* depth of glacial ice at South Pole, *Geophys. Res. Letters*, 27, 2129–2132, 2000.

Proshutinsky, A., R.H. Bourke and F.A. McLaughlin. The role of the Beaufort Gyre in Arctic climate variability: Seasonal to decadal climate scales, *Geophys. Res. Letters*, 29, 15-1 to 15-4, 2002.

Ren, D. and D. Karoly. Comparison of glacier-inferred temperatures with observations and climate model simulations, *Geophys. Res. Letters*, L23710, 2006.

Rothrock, D.A., Y. Yu and G.A. Maykut. Thinning of the Arctic sea-ice cover, *Geophys. Res. Letters*, 26, 3469–3472, 1999.

Scherrer, S.C. and C. Appenzeller. Trends in Swiss Alpine snow days: the role of local- and large-scale climate variability, *Geophys. Res. Letters*, 31, L13215, 2004.

Serreze, M.C., J.A. Maslanic, T.A. Scambos, F. Fetterer, J. Stroeve, K. Knowles, C. Fowler, S. Drobot, R.G. Barry and T.M. Haran. A record minimum Arctic sea ice extent and area in 2002, *Geophys. Res. Letters*, 30, 10-1 to 10-4, 2003.

Sewall, J.O. and L.C. Sloan. Disappearing Arctic sea ice reduces available water in the American west, *Geophys. Res. Letters*, 31, L06209, 2004.

Smith, L.C., Y. Sheng, R.R. Forster, K. Steffen, K.E. Frey and D.E. Alsdorf. Melting of small Arctic ice caps observed from ERS scatterometer time series, *Geophys. Res. Letters*, 30, CRY 2-1 to 2-4, 2003.

Steffen, K., S.V. Nghiem, R. Huff and G. Neumann. The melt anomaly of 2002 on the Greenland ice sheet from active and passive microwave satellite observations, *Geophys. Res. Letters*, 31, L20402, 2004.

Stieglitz, M., S.J. Dery, V.E. Romanovsky and T.E. Osterkamp. The role of snow cover in the warming of arctic permafrost, *Geophys. Res. Letters*, 30, 54-1 to 54-4, 2003.

Stone, J.O., G.A. Balco, D.E. Sugden, J.W. Caffee, L.C. Sass III, S.G. Cowdery and C. Siddoway. Holocene deglaciation of Marie Byrd Land, West Antarctica, *Science*, 299, 99–102, Jan. 3, 2003.

Sturm, M., D.K. Perovich and M.C. Serreze. Meltdown in the north, *Sci. Amer.*, 60–67, Oct. 2003.

Vogel, S.W., S. Tulaczyk, B. Kamb, H. Engelhardt, F.D. Casey, A.E.

Behar, A.L. Lane and I. Joughin. Subglacial conditions during and after stoppage of an Antarctic ice stream: Is reactivation imminent?, *Geophys. Res. Letters*, 32, L022563, 2005.

Winton, M. Amplified Arctic climate change: What does surface albedo feedback have to do with it?, *Geophys. Res. Letters*, 33, L03701, 2006.

Winton, M. Does the Arctic sea ice have a tipping point?, *Geophys. Res. Letters*, 33, L23504, 2006.

Wu, P., R. Wood and P. Stott. Human influence on increasing Arctic river discharges, *Geophys. Res. Letters*, 32, L02703, 2005.

Zhang, T. Influence of the seasonal snow cover on the ground thermal regime: An overview, *Rev. Geophys.*, 43, no. 4, RG4002, Dec. 2005.

General

Blender, R., K. Fraedrich and B. Hunt. Millennial climate variability: GCM-simulation and Greenland ice cores, *Geophys. Res. Letters*, 33, L04710, 2006.

Burroughs, W.J. *Climate Change: A Multidisciplinary Approach*, Cambridge University Press, Cambridge, UK, 2001.

Clark, P.U., R.C. Webb and L.D. Keigwin (eds). Mechanisms of global climate change at millennial time scales, *Amer. Geophys. Union*, Geophys. Monograph 112, Washington, D.C., 1999.

Dai, A., T.M.L. Wigley, G.A. Meehl and W.M. Washington. Effects of stabilizing atmospheric CO_2 on global climate in the next two centuries, *Geophys. Res. Letters*, 28, 4511–4514, 2001.

Flannery, T. *The Weather Makers*, Atlantic Monthly Press, New York, NY, 2005.

Gore, A. *An Inconvenient Truth*, Rondale, Emmaus, PA, 2006.

Hansen, J. Defusing the global warming time bomb, *Sci. Amer.*, 70–77, Mar. 2004.

Hasselmann, K., M. Latif, C. Hooss, C. Azar, O. Edenhofer, C.C. Jaeger, O.M. Johannessen, C. Kemfert, M. Welp and A. Wokaun. The challenge of long-term climate change, *Science*, 302, 1923–1926, Dec. 12, 2003.

Houghton, J. *Global Warming: The Complete Briefing* (3rd edn), Cambridge University Press, 2004.

Karl, T.R. and K.E. Trenberth. Modern global climate change, *Science*, 302, Dec. 5, 2003.

Kattsov, V.M. and P.V. Sporyshev. Timing of global warming in IPCC AR4 AOGCM simulations, *Geophys. Res. Letters*, L23707, 2006.

Kolbert, E. *Field Notes From a Catastrophe*, Bloomsbury Publishing, New York, NY, 2006.

Pierrehumbert, R.T. The hydrologic cycle in deep-time climate problems, *Nature*, 419, 191–198, Sep. 12, 2002.

Snyder, M.A, J.L. Bell and L.C. Sloan. Climate responses to a doubling of atmospheric carbon dioxide for a climatically vulnerable region, *Geophys. Res. Letters*, 29, 9-1 to 9-4, 2002.

Greenhouse Gases

Baird, A.J., C.W. Beckwith, S. Waldron and J.M. Waddington. Ebullition of methane-containing gas bubbles from near-surface *Sphagnum* peat, *Geophys. Res. Letters*, 31, L21505. 2004.

Borges, A.V., S. Djenidi, G. Lacroix, J. Theate, B. Delille and M. Frankignoulle. Atmospheric CO_2 flux from mangrove surrounding waters, *Geophys. Res. Letters*, 30, 12-1 to 12-4, 2003.

Brewer, P.G., C. Paull, E.T. Peltzer, W. Ussler, G. Rehder and G. Friedrich. Measurements of the fate of gas hydrates during transit through the ocean water column, *Geophys. Res. Letters*, 29, 38-1 to 38-4, 2002.

Christensen, T.R., A. Ekberg, L. Strom, M. Mastepanov, N. Panikov, M. Oquist, B.H. Svensson, H. Jykanen, P.J. Marikainen and H. Oskarsson. Factors controlling large scale variations in methane emissions from wetlands, *Geophys. Res. Letters*, 30, 67-1 to 67-4, 2003.

Crowley, T.J. and R.A. Berner. CO_2 and climate change, *Science*, 292, 870–872, May 4, 2001.

Feely, R.A., C.L. Sabine, K. Lee, W. Berelson, J. Kleypas, F.J. Fabry and F.J. Millero. Impact of anthropogenic CO_2 on the $CaCO_3$ system in the oceans, *Science*, 305, 362–366, Jul. 16, 2004.

Ferretti, D.F., J.B. Miller, J.W.C. White, D.M. Etheridge, K.R. Lassey, D.C. Lowe, C.M. MacFarling Meure, M.F. Dreier, C.M. Trudinger, T.D. van Ommen and R.L. Langenfelds. Unexpected changes to the global methane budget over the past 2000 years, *Science*, 309, 1714–1717, Sep. 9, 2005.

Hall, T.M. and F.W. Primeau. Separating the natural and anthropogenic air-sea flux of CO_2: Indian Ocean, *Geophys. Res. Letters*, 31, L23305, 2004.

Heeschen, K.U., A.M. Trehu, R.W. Collier, E. Suess and G. Rehder. Distribution and height of methane bubble plumes on the Cascadia Margin characterized by acoustic imaging, *Geophys. Res. Letters*, 30, 45-1 to 45-4, 2003.

Hinrichs, K.-U., L.R. Hmelo and S.P. Sylva. Molecular fossil record of elevated methane levels in late Pleistocene coastal waters, *Science*, 299, 1214–1217, Feb. 21, 2003.

Hungate, B.A., J.S. Dukes, J.R. Shaw, Y. Luo and C.B. Field. Nitrogen and climate change, *Science*, 302, 1512–1513, Nov. 28, 2003.

Kump, L.R. Reducing uncertainty about carbon dioxide as a climate driver, *Nature*, 419, 188–190, Sep. 12 2002.

Rehder, G., P.W. Brewer, E.T. Peltzer and G. Friederich. Enhanced lifetime of methane bubble streams within the deep ocean, *Geophys. Res. Letters*, 29, 21-1 to 21-4, 2002.

Rex, M., R.J. Salawitch, P. von der Gathen, N.R.P. Harris, M.P. Chipperfield and B. Naujokat. Artic ozone loss and climate change, *Geophys. Res. Letters*, 31, L04116, 2004.

Schmale, O., J. Greinert and G. Rehder. Methane emission from high-intensity marine gas seeps in the Black Sea into the atmosphere, *Geophys. Res. Letters*, 32, L07609, 2005.

Scholze, M., J.O. Kaplan, W. Knorr and M. Heimann. Climate and interannual variability of the atmosphere-biosphere $^{13}CO_2$ flux, *Geophys. Res. Letters*, 30, 69-1 to 69-4, 2003.

Spahni, R., J. Schwander, J. Fluckiger, B. Stauffer, J. Chappellaz and D. Raynaud. The attenuation of fast atmospheric CH_4 variations recorded in polar ice cores, *Geophys. Res. Letters*, 30, 25-1 to 25-4, 2003.

Takahashi, K., T. Nakayama, Y. Matsumi, S. Soloman, T. Gejo, E. Shigemasa and T.J. Wallington. Atmospheric lifetime of SF_5CF_3, *Geophys. Res. Letters*, 29, 7-1 to 7-4, 2002.

Takahashi, T. The fate of industrial carbon dioxide, *Science*, 305, 352–353, Jul. 16, 2004.

Valdes, P.J., D.J. Beerling and C.E. Johnson. The ice age methane budget, *Geophys. Res. Letters*, 32, L02704, 2005.

Voldoire, A. Quantifying the impact of future land-use changes against increases in GHG concentrations, *Geophys. Res. Letters*, 33, L04701, 2006.

Washenfelder, R.A., P.O. Wennberg and G.C. Toon. Tropospheric methane retrieved from ground-based near-IR solar absorption spectra, *Geophys. Res. Letters*, 30, ASC 13-1 to 13-4, 2003.

Wood, W.T., J.F. Gettrust, N.R. Chapman, G.D. Spence and R.D. Hyndman. Decreased stability of methane hydrates in marine sediments owing to phase-boundary roughness, *Nature*, 420, 656–660, Dec. 12, 2002.

Zhang, Y. Methane escape from gas hydrate systems in marine environment, and methane-driven oceanic eruptions, *Geophys. Res. Letters*, 30, 51-1 to 51-4, 2003.

References

Mitigation of Global Warming

Caldeira, K., M.E. Wickett and P.B. Duffy. Depth, radiocarbon, and the effectiveness of direct CO_2 injection as an ocean carbon sequestration strategy, *Geophys. Res. Letters*, 29, 13-1 to 13-4, 2002.

Harvey, L.D.D. Impact of deep-ocean sequestration on the atmosphere CO_2 and on surface-water chemistry, *Geophys. Res. Letters*, 30, 41-1 to 41-4, 2003.

Haugan, P.M. and F. Joos. Metrics to assess the mitigation of global warming by carbon capture and storage in the ocean and in geological reservoirs, *Geophys. Res. Letters*, 31, L18202, 2004.

Hoffert, M.I. and 17 co-authors. Advanced technology paths to global climate stability; Energy for a greenhouse planet, *Science*, 298, 981–987, Nov. 1, 2002.

Lackner, K.S. A guide to CO_2 sequestration, *Science*, 300, 1677–1678, Jun. 13, 2003.

Lal, R. Soil carbon sequestration impact on global climate change and food security, *Science*, 304, 1623–1627, Jun. 11, 2004.

Tromp, T.K., T.-L. Shia, M. Allen, J.M. Eller and Y.L. Yung. Potential environmental impact of a hydrogen economy on the stratosphere, *Science*, 300, 1740–1742, Jun. 13, 2003.

Various authors. Toward a hydrogen economy, *Science*, 305, 962–973, Aug. 13, 2004.

Warwick, N.J., S. Bekki, E.G. Nisbet and J.A. Pyle. Impact of a hydrogen economy on the stratosphere and troposphere studied by 2-D model, *Geophys. Res. Letters*, 31, L05107, 2004.

Winters, J. Carbon underground, *Mech. Engineering*, 46–48, Feb. 2003.

Oceans

Anderson, L and A. Olsen. Air-sea flux of anthropogenic carbon dioxide in the North Atlantic, *Geophys. Res. Letters*, 29, 16-1 to 16-4, 2002.

Antonov, J.I., S. Levitus and T.P. Boyer. Climatological annual cycle of ocean heat content, *Geophys. Res. Letters*, 31, L04304, 2004.

Bertler, N.A.N., P.J. Barrett, P.A. Mayewski, R.L. Fogt, K.J. Kreutz and J. Shulmeister. El Nino suppresses Antarctic warming, *Geophys. Res. Letters*, 31, L15207, 2004.

Bigg. G. *The Oceans and Climate*, Cambridge University Press, 273 pp., 2003.

Caldeira, K. and M.E. Wickett. Anthropogenic carbon and ocean pH, *Nature*, 425, 365, Sep. 25, 2003.

Chan, W.-L and T. Motoi. Response of thermohaline circulation and thermal structure to removal of ice sheets and high atmosphere CO_2 concentration, *Geophys. Res. Letters*, 32, L07601, 2005.

Crowley, T.J., S.K. Baum, K-Y. Kim, G.C. Hegerl and W.T. Hyde. Modeling ocean heat content changes during the last millennium, *Geophys. Res. Letters*, 30, CLM 3–1 to 3–4, 2003.

Delworth, T.L. and K.W. Dixon. Have anthropogenic aerosols delayed a greenhouse gas-induced weakening of the North Atlantic thermohaline circulation?, *Geophys. Res. Letters*, 33, L02606, 2006.

Dong, B.-W. and R.T. Sutton. Adjustment of the coupled ocean-atmosphere system to a sudden change in the thermohaline circulation, *Geophys. Res. Letters*, 29, 18-1 to 18-4, 2002.

Foltz, G.R. and M.J. McPhaden. Unusually warm sea surface temperatures in the tropical North Atlantic during 2005, *Geophys. Res. Letters*, 33, L19703, 2006.

Gillett, N.P., F.W. Zwiers, A.J. Weaver and P.A. Stott. Detection of human influence on sea-level pressure, *Nature*, 422, 292–294, 2003.

Hall, T.M. and F.W. Primeau. Separating the natural and anthropogenic air–sea flux of CO_2: The Indian Ocean, *Geophys. Res. Letters*, 31, L23302, 2004.

Levitus, S., J.I. Antonov, T.P. Boyer and C. Stephens. Warming of the world ocean, *Science*, 287, 2225–2229, Mar. 24, 2000.

Liu, Z., S. Shin, R.S. Webb, W. Lewis and B.L. Otto-Bliesner. Atmospheric CO_2 forcing on glacial thermohaline circulation and climate, *Geophys. Res. Letters*, 32, L02706, 2005.

Meskhidze, N., W.L. Chaneides, A. Nenes and G. Chen. Iron mobilization in mineral dust: Can anthropogenic SO_2 emissions affect ocean productivity?, *Geophys. Res. Letters*, 30, ASC 2-1 to 2-4, 2003.

Raper, C.B. and R.J. Braithwaite. The potential for sea level rise: New estimates from glacier and ice cap area and volume distributions, *Geophys. Res. Letters*, 32, L05502, 2005.

Renssen, H., H. Goosse and T. Fichefet. On the non-linear response of the ocean thermohaline circulation to global deforestation, *Geophys. Res. Letters*, 30, 33-1 to 33-4, 2003.

Seidov, D. and B.J. Haupt. Freshwater teleconnection and ocean thermohaline circulation, *Geophys. Res. Letters*, 30, 62-1 to 62-4, 2003.

Shin, S.-I., Z. Liu, B.L. Otto-Bliesner, J.E. Kutzbach and S.J. Vavus. Southern ocean sea-ice control of the glacial North Atlantic thermohaline circulation, *Geophys. Res. Letters*, 30, 68-1 to 68-4, 2003.

Timmermann, A., F.-F. Jin and M. Collins. Intensification of the annual cycle in the tropical Pacific due to greenhouse warming, *Geophys. Res. Letters*, 31, L12208, 2004.

Thorpe, R.B., J.M. Gregory, T.C. Johns, R.A. Wood and J.F.B. Mitchell. Mechanisms determining the atlantic thermohaline circulation response to greenhouse gas forcing in a non-flux-adjusted coupled climate model, *J. of Climate*, 14, 3102–3116, 2001.

Thresher, R., S.R. Rintoul, J.A. Koslow, C. Weidman, J. Adkins and C. Proctor. Ocean evidence of climate change in southern Australia over the last three centuries, *Geophys. Res. Letters*, 31, L07212, 2004.

Vaughan, D.G. How does the Antarctic ice sheet affect sea level rise?, *Science*, 308, 1877–1878, 2005.

Wadhams, P. and W. Munk. Ocean freshening, sea level rising, sea ice melting, *Geophys. Res. Letters*, 31, L11311, 2004.

Zhang, K., B.C. Douglas and S.P. Leatherman. Global warming and coastal erosion, *Climate Change*, 64, 41–58, 2004.

Paleoclimates

Bond, G., W. Showers, M. Cheseby, R. Lotti, P. Almasi, P. deMenocal, P. Priore, H. Cullen, I. Hajdas and G. Bonani. A pervasive millennial-scale cycle in North Atlantic Holocene and glacial climates, *Science*, 278, 1257–1266, 1997.

Campbell, I.D., C. Campbell, M.J. Apps, N.W. Rutter and A.B.G. Bush. Late Holocene ~1500 yr climatic periodicities and their implications, *Geology*, 26, 271–473, 1998.

Chenoweth, M. Two major volcanic cooling episodes derived from global marine air temperature, AD 1807–1827, *Geophys. Res. Letters*, 28, 2963–2966, Aug. 1, 2001.

Fedorov, A.V., P.S. Dekens, M. McCarthy, A.C. Ravelo, P.B. de Menocal, M. Barreiro, R.C. Pacanowski and S.G. Philander. The Pliocene paradox (Mechanisms for a permanent El Niño), *Science*, 312, 1485–1489, Jun 9, 2006.

Goosse, H., T.J. Crowley, E. Zorita, C.M. Ammann, H. Renssen and E. Driesschaert. Modeling the climate of the last millennium: What causes the differences between simulations?, *Geophys. Res. Letters*, 32, L06710, 2005.

Landais, A., J.M. Barnola, V. Masson-Delmotte, J. Jouzel, J. Chappellaz, N. Caillon, C. Huber, M. Leuenberger and S.J. Johnsen. A continuous record of temperature evolution over a sequence of Dansgaard–Oeschger events during Marine Isotopic Stage 4 (76 to 62 kyr BP), *Geophys. Res. Letters*, 31, L22211, 2004.

Lewis, J.P., A.J. Weaver and M. Eby. Deglaciating the snowball Earth: Sensitivity to surface albedo, *Geophys. Res. Letters*, L23604, 2006.

Lourens, L.J., A. Sluijs, D. Kroon, J.C. Zachos, E. Thomas, U. Rohl, J. Bowles and I. Raffi. Astronomical pacing of late Palaeocene to early Eocene global warming event, *Nature*, 435, 1083–1087, Jun. 23, 2005.

Lyman, J.M, J.K. Willis and G.C. Johnson. Recent cooling in the upper ocean, *Geophys. Res. Letters*, 33, L18604, 2006.

O'Reilly, B.M. and P.W. Readman. Cold-water coral mounds: Evidence for early Holocene climate change and slope failure, *Geophys. Res. Letters*, 31, L07204, 2004.

Perkins, S. Once upon a lake, *Science*, 162, 283, Nov. 2, 2002.

Reason, C.J.C. and M. Rouault. Sea surface temperature variability in the tropical Atlantic Ocean and West African rainfall, *Geophys. Res. Letters*, 33, L21705, 2006.

Renssen, H., H. Goosse, T. Fichefet and J.-M. Campin. The 8.2 kyr BP event simulated by a global atmosphere-sea-ice-ocean model, *Geophys. Res. Letters*, 28, 1567–1570, Apr. 15, 2001.

Schulz, M., W.H. Berger, M. Sarnthein and P.M. Grootes. Amplitude variations of 1470-year climate oscillations during the last 100,000 years linked to fluctuations of continental ice mass, *Geophys. Res. Letters*, 26, 3385–3388, 1999.

Politics

Allen, M. Liability for climate change, *Nature*, 421, 891–892, Feb. 27, 2003.

Gelbspan. R. *The Heat Is On; The Climate Crisis, the Coverup, the Prescription*, Perseus Books, Cambridge, MA, 1998.

King, D.A. Climate change science: Adapt, mitigate, or ignore, *Science*, 303, 176–177, Jan. 9, 2004.

Mooney, C. *The Republican War on Science*, Basic Books, New York, 2005.

Schrope, M. A change of climate for big oil, *Nature*, 411, 516–518, May 31, 2001.

Victor, D.G., J.C. House and S. Joy. A Madisonian approach to climate policy, *Science*, 309, 1820–1821, Sep. 16, 2005.

Wigley, T.M.L. The Kyoto Protocol: CO_2, CH_4 and climate implications, *Geophys. Res. Letters*, 25, 2285–2288, Jul. 1, 1998.

Solar Irradiance

Douglass, D.H. and B.D. Clader. Climate sensitivity of the Earth to solar irradiance, *Geophys. Res. Letters*, 29, 33-1 to 33-4, 2002.

Osborn, T.J. and K.R. Briffa. The real color of climate change?, *Science*, 306, 621–622, Oct. 22, 2004.

Pang, K.D. and K.K. Yau. Ancient observations link changes in sun's brightness to Earth's climate, *EOS*, 83, 481, Oct. 22, 2002.

Philipona, R. and B. Durr. Greenhouse forcing outweighs decreased solar radiation driving rapid temperature rise over land, *Geophys. Res. Letters*, 31, L22208, 2004.

Temperature

Beltrami, H. Climate from borehole data: Energy fluxes and temperatures since 1500, *Geophys. Res. Letters*, 29, 26-1 to 26-4, 2002.

Block, A., K. Keuler and E. Schaller. Impacts of anthropogenic heat on regional climate patterns, *Geophys. Res. Letters*, 31, L12211, 2004.

Karl, T.R., R.W. Knight and B. Baker. The record breaking global temperatures of 1997 and 1998: Evidence for an increase in the rate of global warming?, *Geophys. Res. Letters*, 27, 719–722, 2000.

Nicholls, N. Continued anomalous warming in Australia, *Geophys. Res. Letters*, 30, 23-1 to 23-4, 2003.

Zhai, P. and X. Pan. Trends in temperature extremes during 1951–1999 in China, *Geophys. Res. Letters*, 30, CLM 9-1 to 9-4, 2003.

Weather

Alexander, L.V., S.F.B. Tett and T. Jonsson. Recent observed changes in severe storms over the United Kingdom and Iceland, *Geophys. Res. Letters*, 32, L13704, 2005.

Beniston, M. Warm winter spells in the Swiss Alps: Strong heat waves in a cold season? A study focusing on climate observations at the Saentis high mountain site, *Geophys. Res. Letters*, 32, L01812, 2005.

Fowler, H.J. and C.G. Kilsby. Implications of changes in seasonal and annual extreme rainfall, *Geophys. Res. Letters*, 30, 53-1 to 53-4, 2003.

Gao, X., J.S. Pal, and F. Giogi. Projected changes in mean and extreme precipitation over the Mediterranean region from a high resolution double nested RCM simulation, *Geophys. Res. Letters*, 33, L03706, 2006.

McHugh, M.J. Multi-model trends in East African rainfall associated with increased CO_2. *Geophys. Res. Letters*, 32, L01707, 2005.

Pal, J.S., F. Giorgi and X. Bi. Consistency of recent European summer precipitation trends and extremes with future regional climate projections, *Geophys. Res. Letters*, 31, L13202, 2004.

Stott, P.A., D.A. Stone and M.R. Allen. Human contribution to the European heat wave of 2003, *Nature*, 432, 610–614, Dec. 2, 2004.

Xie, L., T. Yan, and L. Pietrafesa. The effect of Atlantic sea surface temperature dipole mode on hurricanes: Implications for the 2004 Atlantic hurricane season, *Geophys. Res. Letters*, 32, L03701, 2005.

INDEX

Printed in Singapore